Forskolin

T. Pullaiah

Forskolin

Natural Sources, Pharmacology and Biotechnology

Springer

T. Pullaiah
Department of Botany
Sri Krishnadevaraya University
Anantapur, Andhra Pradesh, India

ISBN 978-981-19-6520-3 ISBN 978-981-19-6521-0 (eBook)
https://doi.org/10.1007/978-981-19-6521-0

This Springer imprint is published by the registered company Springer Nature Singapore Pte Ltd.
The registered company address is: 152 Beach Road, #21-01/04 Gateway East, Singapore 189721, Singapore

Preface

Forskolin, a naturally occurring labdane diterpenoid, is useful in the treatment of obesity, diabetes, heart diseases, cancer, asthma, glaucoma and hypertension. *Coleus forskohlii* is the only source of Forskolin. We were working on different aspects of micropropagation, in vitro production of secondary metabolites for the last several years. We found that by 2006 itself more than 17,000 papers using forskolin as a pharmacological tool have been published in scientific literature. We also found that there is no comprehensive account on Forskolin and *Coleus forskohlii*. This book is an answer to this.

I proposed this book with Dr. Mohammad Majeed, of Sami Sabinsa Corporation, pioneer in Coleus research. A visionary, he migrated to USA with 8 US Dollars and established a company, based on Coleus and Curcumin, which is now worth Rupees one thousand crores. First he readily agreed but due to his busy business work he opted out and I had to take whole load in editing this book. Some of his colleagues initially gave their helping hand. I thank them for the same. My students came to my rescue to complete the job.

I thank all the journals, who gave me permission to reproduce some of the figures from their publication. I thank Springer Nature, especially Shri Naren Agarwal and Momoko Asawa for their support in bringing this book to a shape.

Anantapur, Andhra Pradesh, India T. Pullaiah

Contents

About the Author

T. Pullaiah is a former Professor at the Department of Botany at Sri Krishnadevaraya University in Andhra Pradesh, India, where he has taught for more than 35 years. He has held several positions at the university. He was President of Indian Botanical Society (2014), President of the Indian Association for Angiosperm Taxonomy (2013) and Fellow of Andhra Pradesh Akademi of Sciences. Under his guidance 54 students obtained their doctoral degrees. He has authored 70 books, edited 40 books, and published over 340 research papers, including reviews and book chapters. He was also a member of Species Survival Commission of the International Union for Conservation of Nature (IUCN). Professor Pullaiah received his PhD from Andhra University, India, attended Moscow State University, Russia, and worked as Post-Doctoral Fellow during 1976–1978.

Coleus forskohlii—A Multipurpose Medicinal Plant

1

Abstract

Forskolin is a naturally occurring labdane diterpenoid, derived from the root of *Coleus forskohlii*. Forskolin, a diterpenoid the compound of our interest, is useful in preventing the clotting of platelets, in reducing intraocular pressure in cases of glaucoma, as a hypotensive, and as an aid to nerve regeneration following trauma. *Coleus forskohlii* is the only species to contain forskolin.

Keywords

Coleus forskohlii · *Forskolin* · *Labdane diterpenoid*

1.1 Introduction

For millennia, humankind has used plants and plant-derived drugs for the treatment or alleviation of many types of diseases. It is estimated that one-fourth of prescription drugs contain at least one chemical originally identified and extracted from a plant.

The types of plant constituents used as drugs are classified as "Secondary metabolites" which are biosynthetically derived from primary metabolites, but are more limited in distribution in the plant kingdom, usually being restricted to a particular taxonomic group. Although they have no obvious roles in a plant's, primary or "main stream" metabolism, secondary compounds often play ecologically significant roles in how plants deal with their environment and are therefore important in their ultimate survival. Since plant secondary metabolites serve basically to combat infectious diseases.

In terms of cellular economy, secondary metabolites are generally metabolically expensive to produce and accumulate and are therefore frequently present in plants in much smaller quantities than primary metabolites. In addition, secondary metabolites tend to be biosynthesized in specialized cell types and at distinct

T. Pullaiah, *Forskolin*, https://doi.org/10.1007/978-981-19-6521-0_1

development stages, making their extraction, isolation, and purification difficult. Secondary metabolites that are used commercially as biologically active compounds are generally higher value-low volume products. Thus, secondary metabolites can be considered as special materials or fine chemicals (Balandrin and Klocke 1988).

Forskolin (7-β-acetoxy-8, 13-epoxy-1α, 6 β, 9α-trihydroxy-labd-14-ene-11 one), a naturally occurring labdane diterpenoid, derived from the tuberous roots of *Coleus forskohlii* (Willd.) Briq. (Syn.: *Plectranthus barbatus* Andrews) an ancient root drug of Ayurveda, the codified Indian medicine. *Coleus* belongs to the family Lamiaceae. *C. forskohlii*, cited in ancient Hindu and Ayurvedic texts, is gaining its importance globally. Forskolin is useful in preventing the clotting of platelets, in reducing intraocular pressure in cases of glaucoma, as a hypotensive, and as an aid to nerve regeneration following trauma. Forskolin is useful in the treatment of obesity, diabetes, heart diseases, cancer, asthma, glaucoma, and hypertension (Khatun et al. 2011).

Coleus forskohlii is the only species to contain forskolin and enjoys a unique position in the field of chemotaxonomy. Shah et al. (1980) searched for alternative and/or superior sources of forskolin within the family Lamiaceae in roots and shoots of six species belonging to the genus *Coleus* and six other species belonging to the allied genera and indicated that only *C. forskohlii* contained forskolin.

Chemical assay of roots and shoots of 24 species of the following genera (value in parentheses refers to the number of species) viz. *Anisochilus* (4) *Hyptis* (1), *Lavandula* (1) *Basilicum* (1), *Ocimum* (1), *Orthosiphon* (3) *Anisomeles* (1), *Colebrookea* (1) *Dysophylla* (1) *Gomphostemma* (1), *Leucoceptrum* (1), *Leonurus* (1), *Nepeta* (1), *Pogostemon* (2), *Salvia* (3), and *Stachys* (1) in the family Lamiaceae also confirmed the exclusive presence of forskolin in *C. forskohlii*.

In nature, *C. forskohlii* exhibits several fascinating morphologic variations in the plants collected from different ecogeographic regions. Shah (1989) has given a comprehensive account of morphological, histological, cytological, palynological, phonological, and agronomical variations in 10 natural and two cultivar ecotypes located at 12 diverse ecogeographic sites. Striking morphological variations were observed in the roots, growth habit, shoot height, number of branches, pattern of branching, leaf morphology, floral bracts, inflorescence length, and indumentum. Vishwakarma et al. (1988) reported variations in forskolin content from 0.01 to 0.44% in 6-month-old dry roots. Scientists at Hoechst India Limited in their search for forskolin-rich strain, assayed 102 root samples of single clones and of populations collected from natural sources from different ecogeographic regions in India and more than 12 cultivar root samples from Gujarat, Maharashtra, and Belgaum farms in Karnataka between the years 1982 and 1986 for forskolin content. Forskolin content in these samples ranged between 0.066% and 0.58% by dry weight.

Forskolin, exclusively found in the root of *C. forskohlii*, has been used in traditional Indian Ayurvedic and Southeast Asian medicine since ancient times (Kanne et al. 2015). *C. forskohlii* is commonly found in Nepal, Burma, Thailand, and India. India is a leading exporter of *C. forskohlii* extracts and their products to various countries (Bhowal and Mehta 2017). By 2006 itself more than 17,000 papers using forskolin as a pharmacological tool have been published in scientific literature (Staudinger et al. 2006).

References

Balandrin MF, Klocke JA (1988) Medicinal, aromatic and industrial materials from plants. In: Bajaj YPS (ed) Biotechnology in agriculture and forestry 4. Medicinal and aromatic plants I. Springer-Verlag, Berlin, pp 3–36

Bhowal M, Mehta DM (2017) *Coleus forskohlii*: phytochemical and pharmacological profile. Int J Pharm Sci Res 8(9):3599–3618

Kanne H, Burte NP, Prasanna V, Gujjula R (2015) Extraction and elemental analysis of *Coleus forskohlii* extract. Pharm Res 7:237–241

Khatun S, Cakilcioglu U, Chatterjee NC (2011) Pharmacognostic value of leaf anatomy and trichome morphology for identification of forskolin in a novel medicinal plant *Coleus forskohlii*. Biol Divers Conserv 4(2):165–171

Shah V, Bhat SV, Bajura BS, Dornauer H, DeSouza NJ (1980) The occurrence of forskolin in Labiatae. Planta Med 39:183–185

Shah VC (1989) Biosystcmatic studies on *Coleus barbatus* (Andr.) Benth. Ph.D thesis. University of Bombay, Bombay, India

Staudinger JL, Ding X, Lichti K (2006) Pregnane X receptor and natural products: beyond drug-drug interactions. Expert Opin Drug Metab Toxicol 2:847–857

Vishwakarma RA, Tyagi BR, Ahmed B, Husain A (1988) Variation in forskolin content in the roots of *Coleus forskohlii*. Planta Med 54(5):471–472

2 Botany of *Coleus forskohlii*

Abstract

The genus *Coleus* earlier consisted of about 150 species, but many of these have now been transferred to other genera. The species of the *Coleus* are perennial, much branched and aromatic herbs. They grow well on dry hill slopes exposed to sunlight from 300 to 1800 m altitude. The Plant List includes 56 scientific plant names of species rank for the genus *Coleus*; of these only two species, i.e., *Coleus decurrens* Gürke, and *Coleus forskohlii* (Willd.) Briq. are accepted species names. *Coleus forskohlii* is an aromatic, perennial, and herbaceous species grown in tropical and temperate regions of the World. The name *forskohlii* was given to commemorate the Finnish Botanist, Forskel. In this chapter, Botany of *Coleus*, including chromosome number, genetic improvement, and mutation breeding are given.

Keywords

Coleus forskohlii · Taxonomy

2.1 Introduction

The name *Coleus* is derived from Greek word *Koleus* means sheathing around the style. The genus *Coleus* earlier consisted of about 150 species, but many of these have been transferred to other genera. The species of the *Coleus* are perennial, much branched and aromatic herbs. The species are distributed in Asia, tropical Africa, Australia, the Philippines, East Indies, and Malay Archipelago. More than 500 varieties of *Coleus* are cultivated for their colorful variegated leaves (Uphof 1959). They grow well on dry hill slopes exposed to sunlight from 300 to 1800 m altitude. The species of the genus have branched quadrangular stem with hairy nature. Roots are thick, tuberous, or fasciculate, fusiform with aromatic smell. The flowers are pale bluish to laver color, corolla bi-lipped, and lower lobes are elongated

T. Pullaiah, *Forskolin*, https://doi.org/10.1007/978-981-19-6521-0_2

and concave. The leaves and roots are with characteristic odor. The major chemical constituents of the *Coleus* species are volatile oils, glycosides, flavonoids, and phenolic components (Soni Himesh and Singhai 2012). The species of the *Coleus* have been reported to use to cure different human ailments in different cultures.

2.2 Botany of *Coleus forskohlii*

The genus *Coleus* was described by de Loureiro in 1790 for the species *C. amboinicus* in his Flora Cochinchinensis. The character conveyed by the generic name is also the character of circumscription of the genus and segregates genus *Coleus* from its allied earlier published genus *Plectranthus* having free stamens. Presently, there is no unanimity regarding generic status of *Coleus* Taxonomists (Morton 1962) working on the African flora observing varying degrees of fusion of the stamens in certain populations of *Coleus* disagreed to accept the character of monadelphous stamens as a reliable character and merged all the *Coleus* spp. under a larger generic concept of *Plectranthus*. Species of *Coleus* growing over their range of distribution in the Indian Subcontinent have been examined and found consistently to show distinct character of connate filaments.

Today, there are more than 500 varieties of *Coleus* in cultivation all over the world. *Coleus* plants are very colorful and can be grown indoors as well as outdoors.

The Plant List includes 56 scientific plant names of species ranked for the genus *Coleus*. Of these only two species—*Coleus decurrens* Gürke and *Coleus forskohlii* (Willd.) Briq. are accepted species names (Anon n.d. http://www.theplantlist.org/1.1/browse/A/Lamiaceae/Coleus/).

These are:

Coleus anamallayensis Bedd.
Coleus aulihanensis Schweinf. & Volkens.
Coleus brazzavillensis A. Chev.
Coleus bullulatus Briq.
Coleus casamancicus A. Chev. ex Hutch. & Dalziel.
Coleus chevalieri Briq.
Coleus clandulosus Hook. f.
***Coleus decurrens* Gürke**—accepted.
Coleus denudatus (A. Chev. ex Hutch. & Dalziel) Robyns.
Coleus doba Hochst. ex Chiov.
Coleus × *eureka* auct.
***Coleus forskohlii* (Willd.) Briq**.—accepted.
Coleus gazensis S. Moore.
Coleus gibsonii Verl.
Coleus huberi Regel.
Coleus kisanfuensis De Wild.
Coleus lageniocalyx Briq.
Coleus lateriticola A. Chev.

Coleus leucophyllus Baker.
Coleus luengerensis Gürke.
Coleus macrostachys Benth.
Coleus × *marshallii* T. Moore.
Coleus matopensis S. Moore.
Coleus mechowianus Briq.
Coleus montanus Hochst. ex Ces.
Coleus montanus Gürke.
Coleus mucosus Hayata.
Coleus myrianthellus Briq.
Coleus nervosus Briq.
Coleus omahekense Dinter.
Coleus orbicularis Baker.
Coleus osmirrhizon Elliot.
Coleus palliolatus S. Moore.
Coleus parensis Gürke.
Coleus parvifolius Benth.
Coleus petiolatissimus Briq.
Coleus petrophilus Gürke.
Coleus poggeanus Briq.
Coleus polyanthus S. Moore.
Coleus preussii Gürke.
Coleus saxicola Gürke.
Coleus subbaraoi Kumari & Malathi.
Coleus subscandens Gürke.
Coleus trichophorus Briq.
Coleus × *tryonii* auct.
Coleus ulugurensis Gürke.
Coleus × *veitchii* Dombrain.
Coleus viridis Briq.
Coleus wugensis Gürke.
Coleus xanthanthus C.Y. Wu & Y.C. Huang.

The Plant List treats *Coleus forskohlii* (Willd.) Briq. and *Plectranthus barbatus* Andrews as separate species while most others treat them as conspecific. In this book, we treat them as conspecific. Finnish botanist, Peter Forsskål was commemorated with this species name *forskohlii*.

Coleus species are native to tropical and subtropical regions of Africa, Australia, the East Indies, the Malay Archipelago, and the Philippines (Lebowitz 1985). In India, *C. forskohlii* is observed to grow over a wide geographic range between the latitudes 8° and 31°N, in subtropical and warm temperature climates (Mukherjee 1940). *C. forskohlii* is an herbaceous, pubescent weakly aromatic species with annual stems and perennial rootstock. Roots are polymorphic tuberous, semi-tuberous, and non-tuberous (fibrous). Tuberous roots are one to several, succulent but hard, tortuous or straight, short and stout, or long and slender. The cultivated

Fig. 2.1 *Coelus forskohlii* (Source: https://www.indiamart.com/proddetail/coleus-forskohlii-tablets-10885766933.html)

types are fleshy and succulent spindle shaped or fusiform generally long, not so stout, several, and radially spread. Root tubers in both kinds have white or orangish pink flesh, have bitter aftertaste, and are aromatic.

Coleus forskohlii (Willd.) Briq. in Engl. & Prantl, Pflanzenfam. 4(3a): 359. 1897. *Plectranthus forskohlii* Willd., Sp. Pl. 3: 169. 1800, non Vahl 1790. *Plectranthus barbatus* Andrews, Bot. Repos. 9: t. 594. 1810; Codd., Bothalia 11: 394. 1975. *Coleus barbatus* (Andr.) Benth. ex G. Don in J.C. Loudon, Hort. Brit. 483. 1924; Wall., Pl. Asiat. Rar. 2: 15. 1830–1831; Hook. f. Fl. Brit. India 4: 625. 1885; Gamble, Fl. Madras 2: 1129. 1924; Paton et al., Phytokeys 129: 23. 2019.

Perennial aromatic herb, 30–60 cm high; stem decumbent and ascending, stout, villous; roots thick, and fleshy. Roots tuberous. Leaves fleshy, ovate or obovate 2.5–10 × 1.3–3.5 cm, base narrowed, margin crenate, apex obtuse, pubescent on both sides, gland-dotted below. Flowers in 6–10 flowered whorls, in densely hairy, spiciform racemes; bracts conspicuous, broadly ovate, mucronate, pubescent and ciliate, caducous; calyx hairy, 5 mm long, glandular, throat villous, lower lip with four teeth, subulate-sub aristate, upper lip ovate, acuminate; corolla purple or pale blue, bent at a right angle, upper lip short, broadly ovate, sub-equally 4-lobed, lower lip boat-shaped, joined to the tube by a narrow neck, villous without; stamens 4, didynamous, filaments connate at base into a sheath, anthers blue, with red glands on the lower surface, style slender, long, stigma bifid. Nutlets 4, smooth, and dark brown (Khan et al. 2012) (Figs. 2.1, 2.2, and 2.3).

Occasional in laterite and rocky areas.

Fig. 2.2 Coleus crop (Source https://www.aromediherbs.com/coleus-forskohlii.html)

Fig. 2.3 *Coleus forskohlii* roots (Source: https://tnau.ac.in/hcri-coimbatore/department-of-ma-technologies-developed/)

Fl. & Fr.: October–December.

World Distribution. Eritrea to East and Central Africa, Arabian Peninsula, Indian Subcontinent to SC. China, Myanmar, Thailand. Widely cultivated.

In nature, *C. forskohlii* exhibits several fascinating morphologic variations in the plants collected from different ecogeographic regions. Shah (1989) has given a comprehensive account of morphological, histological, cytological, palynological, phonological, and agronomical variations in 10 natural and two cultivar ecotypes located at 12 diverse ecogeographic sites. Striking morphological variations were observed in the roots, growth habit, shoot height, number of branches, pattern of branching, leaf morphology, floral bracts, inflorescence length, and indumentum. Root morphology differs with populations as they may be fibrous, tuberous, non-tuberous, or semi-tuberous in nature. Tubers and leaves have somewhat different odors. Growth habit of *C. forskohlii* plant is usually variable: decumbent, procumbent, or erect. Vishwakarma et al. (1988) reported variations in forskolin content from 0.01% to 0.44% in 6-month-old dry roots. Scientists at Hoechst India Limited in their search for forskolin-rich strain, assayed 102 root samples of single clones and of populations collected from natural sources from different ecogeographic regions in India and more than 12 cultivar root samples from Gujarat, Maharashtra, and Belgaum farms in Karnataka between the years 1982 and 1986 for forskolin content. Forskolin content in these samples ranged between 0.066% and 0.58% by dry weight.

Abraham et al. (1988) observed pedicillate cytoplasmic vesicles containing yellowish to reddish brown substance storing secondary metabolites in freehand sections of fibrous and tuberous roots of *Coleus forskohlii* collected from Tamil Nadu and Kerala. The vesicles were observed to be common in cork cells in vascular supply to the rootlet and at the periphery of the nematode-infested tissue. Concentration of the vesicles around the infested zone suggests their protective role. The vesicles being exclusive to *C. forskohlii* assume diagnostic importance.

Coleus forskohlii is the only species to contain forskolin and enjoys a unique position in the field of chemotaxonomy. Shah et al. (1980) searched for alternative and/or superior sources of forskolin within the family Lamiaceae in roots and shoots of six species belonging to the genus *Coleus* and six other species belonging to the allied genus and indicated that only *C. forskohlii* contained forskolin.

2.3 Chromosome Number

C. forskohlii is diploid with chromosome number $n = 14$ (Reddy 1952). However, Riley and Hoff (1961) reported that South African *C. forskohlii* is diploid with basic chromosome number $n = 16$. Bir and Saggoo (1982, 1985) reported that Central Indian collections have basic chromosome number of $n = 17$, while South Indian collections have chromosome number $n = 15$. They concluded that variability in base numbers could be due to aneuploidy at generic level which ultimately leads to morphological variations. Shah (1989) reported that populations from different ecogeographic areas vary greatly in their morphology.

2.4 Genetic Improvement

Medicinal Plant Division of Indian Institute of Horticultural Research at Bangalore, worked on polyploidy and hybridization studies from 1989 to 1994. A total of 34 varieties have been developed which include 11 autotetraploids of diploids, 11 intervarietal hybrids, 11 autotetraploids of the hybrids, and one grafted variety. An autotetraploid of the superior variety registered an increase of 23.1% in forskolin content, 1.3% over its diploid progenitor in pot studies. This tetraploid, using diploid progenitor as the check variety, was evaluated in a multilocational trial at four sites—three in Tamil Nadu and one in Karnataka for forskolin accumulation and root tuber yield. Percentage increases of forskolin in the tetraploid varied from site to site, being 2.3, 15, 17.7, and 51%. The root tuber yield, however, was decreased by

47.8, 46.1, 14, and 36.7% at the respective sites, indicating unstable growth characteristics of the autotetraploid and thereby making the autotetraploid unsuitable for commercialization. The diploid cultivar thus continues to be the variety of choice by the farmers (Hegde and Krishna 1991).

2.5 Mutation Breeding

Srinvasappa et al. (2010) obtained high yielding type of *C. forskohlii* through mutation breeding. Remarkable variations were observed in the leaf shape, plant height, leaf color, tuber yield, and forskolin content. The mutant M3RC-2-3 recorded the highest tuber yield (168.64 g/plant) and forskolin content (1.74%), amounting to 155.88% increase over the control K-8 (0.68%). The other three mutants M3UC-5-2, M3RC-1-4, and M3UC-1-3 closely followed the M3RC-2-3 with regard to dry tuber yield per plant (139.48, 122.98, and 141.00 g, respectively), and forskolin content (1.63%, 1.69%, and 1.51%, respectively), resulting in 139.71, 148.53, and 122.06% increase over the control.

In vitro mutation in *C. forskohlii* was carried with shoot tip and callus explants by Velmurugan et al. (2008). Among the mutagenic treatments, the highest forskolin content (0.80%) was observed in 25Gy gamma rays +750 μM EMS. However, the callus-derived in vitro mutants in TC_1M_1 generation produced tallest plants at 120 and 180 DAP, respectively, in 5Gy gamma rays +175 μM EMS treatment. The gradual increase in dose of mutagen expressed a reduction in number of tuber plant-1, length of tuber, fresh and dry tuber yield plant-1. The treatment 1.50kR gamma rays+200 μM EMS produced a 188.89% increase over the control for the forskolin content

References

Abraham Z, Srivastava AK, Bagchi GD (1988) Cytoplasmic vesicles containing secondary metabolites in the root of *Coleus forskohlii* (Willd.) Briq. Curr Sci 57:1377–1379

Anon. (n.d.). http://www.theplantlist.org/tpl1.1/record/kew-157914

Bir SS, Saggoo MIS (1982) Cytology of some members of Labiatae from Central India. Proc Natl Acad Sci India Section B (Biol Sci) 52:107–112

Bir SS, Saggoo MIS (1985) Cytological studies on members of family Labiatae from Kodaikanal and adjoining areas (South India). Proc Indian Acad Sci (Plant Sci) 94:619–626

Hegde L, Krishna R (1991) Varietal response to chromosomal doubling in *Coleus forskohlii* Briq. Trop Agric Res 3:77–89

Khan BA, Akhtar N, Anwar M, Mahmood T, Khan H, Hussain I, Khan KA (2012) Botanical description of *Coleus forskohlii:* a review. Med Plant Res 6:4832–4835

Lebowitz RJ (1985) The genetics and breeding of *Coleus*. Plant Breed Rev 3:343–360

Morton JK (1962) Cytotaxonomic studies on the West African Labiatae. Bot J Linn Soc 58 (372):231–283

Mukherjee SK (1940) A revision of the Labiatae of the Indian Empire. Rec Bot Surv India 14 (1):224–225

Paton AJ, Mwanyambo M, Govaerts RHA, Smitha K, Sudde S, Phillipson RB, Wilson TC, Foster PI, Culham A (2019) Nomenclatural changes in *Coleus* and *Plectranthus* (Lamiaceae): a tale of more than two genera. Phytokeys 129:1–158. https://doi.org/10.3897/phytokeys.129.34988

Reddy NS (1952) Chromosome numbers in *Coleus*. J Hered 43:233–237

Riley HP, Hoff VJ (1961) Chromosome studies in some south African dicotyledons. Can J Genet Cytol 3:260–271

Shah VC (1989) Biosystematic studies on *Coleus barbatus* (Andr) Benth. PhD thesis. University of Bombay, Bombay

Shah V, Bhat SV, Bajwa BS, Dornauer H, de Souza NJ (1980) The occurrence of forskolin in the Labiatae. Planta Med 39:183–185

Soni Himesh, Singhai AK (2012) Recent updates on the genus *Coleus*: a review. Asian J Pharm Clin Res 5:12–17

Srinvasappa KN, Kathiresan C, Farooqi AA, Vasundhara M, Nutan D, Raju B (2010) Crop improvement studies in *Coleus forskohlii* Briq. through mutation breeding. Biomed 5(2):90–97

Uphof JCT (1959) Dictionary of economic plants 191. Wheldon and Whesley, Codicote

Velmurugan M, Rajamani K, Paramaguru P, Kannan Bapu JR, Gnanam R (2008) Induction of variation in *Coleus forskohlii* Briq. Through in vitro mutation technique. International Symposium on induced mutations in plants, Vienna, Austria. Book of Abstracts

Vishwakarma RA, Tyagi BR, Ahmed B, Hussain A (1988) Variation in forskolin content in the roots of *Coleus forskohlii*. Planta Med 54:471–472

3 Traditional Medicinal Uses and Pharmacognosy of *Coleus forskohlii*

Abstract

Coleus forskohlii is an aromatic medicinal plant of the family Lamiaceae. In traditional medicine, plant parts have been used to treat different human ailments throughout the world. In folk medicine, it has been reported as a remedy for laxative, gastric problems, wounds, ringworms, and *Candida* infections. Pharmacognostical analysis of leaf parts of *Coleus forskohlii* indicated the presence of multicellular, uniseriate glandular trichomes as a diagnostic feature. Outer cortex of stem sections showed the presence of 3–4 layers of collenchyma cells. Stelar region consists of 7–9 conjoint, collateral, and open type of vascular bundles arranged in ring-like structure. Root anatomy showed the presence of cork, phellem, phelloderm, phloem, primary and secondary xylem. Leaf samples showed higher physicochemical characters than stem and root samples.

Keywords

Coleus forskohlii · Traditional uses · Ethnobotany · Pharmacognosy

3.1 Introduction

Coleus forskohlii is an aromatic medicinal plant of the family Lamiaceae. In traditional medicine, the plant parts have been used to treat different human ailments throughout the world. It is commonly used in different countries to treat various human diseases. *C. forskohlii* has been used in traditional Indian Ayurvedic medicine, under the name "Makandi" or "Mayani," and Southeast Asian medicine since ancient times (Kanne et al. 2015). The local people of Egypt and Africa, the leaf is used as diuretic, emmenagogue, and as an expectorant, whereas the same part is used to cure stomach and intestinal disorders in Brazil (Valdes et al. 1987). In India, the leaf is used as a condiment and the tubers are used to prepare pickles. In Ayurveda system of medicine, the plant has been used as a remedy for several diseases such as

T. Pullaiah, *Forskolin*, https://doi.org/10.1007/978-981-19-6521-0_3

respiratory disorders, asthma, bronchitis, heart diseases, burning sensation, constipation, insomnia, abdominal colic, epilepsy, convulsions, and angina (Ammon and Muller 1985). The tuberous root parts are used as a medicament for worms and to relieve burning sensation of festering boils. Root tuber extract along with mustard oil is applied externally on skin infections and eczema. It is also reported to have therapeutic value in the veterinary field (De Souza and Shah 1988). Forskolin, the bioactive component of *C. forskohlii,* is used in cosmetics to reduce graying of hair. The root tubers contain essential oil with delicate and attractive odor (Misra et al. 1994). With its spicy aroma, the essential oil is used as flavoring agent in the food industry (Chowdhary and Sharma 1998).

Coleus forskohlii (Syn.: *Coleus barbatus*) is popularly used in Brazil for the healing of liver and stomach diseases (Fischman et al. 1991).

3.2 Uses in Traditional Systems of Medicine

Coleus forskohlii is considered to be the wild ancestor of all the tuber varieties, known as Kaffir Potatoes. It has several ethnomedicinal uses, which have been transmitted by word of mouth from generation to generation. Interestingly, the roots of the plant have a long history of food use in India in the form of pickle/ condiment. According to Ayurveda, it has been used to ease pain, support healthy inflammation response, hemorrhoids, help manage cough, worms, skin-related problems, ascites (fluid retention), external ulcers, abdominal pain, low appetite, urine retention, and constipation (https://forslean.com/forslean/overview/).

Coleus showed therapeutic benefits in curing psoriasis, asthma, angina, and cancer diseases (Thorne Research 2006). Forskolin isolated from *C. forskohlii* in combination with hydroxycitric acid is used to control body fat (Gupta 2004). Essential oil isolated from *C. forskohlii* found to be active against skin disease-causing bacterial species (Majeed and Prakash 2007).

3.3 Ethnobotanical Uses

In folk medicine of Kenya *C. forskohlii* is used to cure stomach pain, nausea, and as a laxative (Lukhoba et al. 2006; Johns et al. 1990; Hamill et al. 2003). In Brazil ethnomedicine it is used to treat intestinal spasms and gastric problems (Camara et al. 2003). In the Democratic Republic of Congo and Kenya, the plant is used as a remedy for wounds and ringworms (Chifundera 2001; Githinji and Kokwaro 1993). In Tanzanian folk medicine, the leaf juice made from fresh leaves of *C. forskohlii* (Syn.: *C. barbatus*) is applied as external and gargle two times to cure Candida infection (Runyoro et al. 2006).

C. forskohlii has been used for centuries in Indian Ayurvedic traditional medicine as well as in the folk medicine of Brazil, tropical Africa, and China for the treatment of various diseases (Table 3.2). In addition, *C. forskohlii* is used to alleviate fever in East Africa and India, as a children's tonic, and also as an emetic utilized by the

Samburu of Kenya for strength. In Uganda, the plant is used to treat spiritual ailments. In Africa, the plant is applied in ethnoveterinary medicine, for instance in Kenya, it is used to treat Coast Fever in cattle *C. forskohlii* is used against snakebites in India, Gabon and Kenya, and as an insecticide to protect grain stores (Alabashi and Melzig, 2010).

Lukhoba et al. (2006) reported that *C. forskohlii* is planted as an ornamental and as a hedge, fence, or boundary marker as well as soil improver for growing grains such as cowpeas, green grams, and maize; it is also planted on the hillsides to prevent soil erosion and is used for making manure. The leaves of *C. forskohlii* are cooked as a vegetable in Kenya and Yemen; it is fed to sheep, goats, and cattle. In Kenya, the soft velvety leaves are used as sanitary tissue to clean milk guards and both leaves and stems are used to hasten the ripening of bananas (Alabashi and Melzig, 2010) (Table 3.1).

3.3.1 Digestive System

In India, *C. forskohlii* is used to treat abdominal colic (Dubey et al. 1981). It is also used to cure stomachache and as a purgative in Kenya and for treating nausea in Southern Uganda (Lukhoba et al. 2006; Matu and van Staden 2003). In Brazil, it is used as a substitute for boldo (*Peumus boldus*) to treat gastric disturbances (e.g., gastritis and intestinal spasms) and hepatic disorders (Lukhoba et al. 2006; Schultz et al. 2007; Kelecom 1983; Camara et al. 2003), teeth and gum disorders (Lukhoba et al. 2006).

3.3.2 Nervous System

In Asia, it is used for treating insomnia, convulsion (Lukhoba et al. 2006; Dubey et al. 1981), and against dizziness and fluster (Yang et al. 2006). In Tanzania, *C. forskohlii* is used for treating psychiatric problems (Lukhoba et al. 2006).

3.3.3 Skin

In East Africa (Kenya, Congo), *C. forskohlii* is used to treat wounds and ringworms, to reduce swelling on bruises and as a bath for babies with measles (Lukhoba et al. 2006; Matu and van Staden 2003).

3.3.4 Respiratory System

C. forskohlii is used to cure asthma, bronchitis, cold, cough, pneumonia (Lukhoba et al. 2006; Jin and He 1998; Yang et al. 2006; Muhayimana et al. 1998), and general respiratory ailments (Dubey et al. 1981; Lukhoba et al. 2006; Matu and van Staden 2003).

Table 3.1 Ethnobotanical uses of *Coleus forskohlii* (Syn.: *Plectranthus barbatus*) (Source: Alabashi and Melzig, 2010, with permission)

Digestive system	**Respiratory system**	**Cardiovascular system**
In India for abdominal colic (Dubey et al. 1981). For stomachache and as purgative in Kenya and for nausea in Southern Uganda (Lukhoba et al. 2006; Matu and van Staden 2003). In Brazil, as a substitute for boldo (*Peumus boldus*) to treat gastric disturbances (e.g., gastritis and intestinal spasms) and hepatic disorders (Lukhoba et al. 2006; Schultz et al. 2007; Kelecom 1983; Camara et al. 2003). Teeth and gum disorders (Lukhoba et al. 2006)	Asthma, bronchitis, cold, cough, and pneumonia (Lukhoba et al. 2006; Jin and He 1998; Yang et al. 2006; Muhayimana et al. 1998). General respiratory ailments (Lukhoba et al. 2006; Dubey et al. 1981; Matu and van Staden 2003)	Angina, hemorrhage, and hypertension (Lukhoba et al. 2006; Dubey et al. 1981)
Nervous system	**Pain, inflammation, musculoskeletal**	**Sensory**
In Asia, for insomnia, convulsion (Lukhoba et al. 2006; Dubey et al. 1981) and against dizziness and fluster (Yang et al. 2006). In Tanzania, for psychiatric problems (Lukhoba et al. 2006)	Inflammation, abdominal and spasmodic pain, and painful micturition (Lukhoba et al. 2006; Dubey et al. 1981). Muscular, generalized pain, stiff neck, backache, bone dislocation, and rheumatism (Lukhoba et al. 2006)	For conjunctivitis in Congo and earache in Kenya (Lukhoba et al. 2006)
Skin	**Metabolic and endocrine system**	**Infection**
In East Africa (Kenya, Congo), for wounds and ringworms, to reduce swelling on bruises and as a bath for babies with measles (Lukhoba et al. 2006; Matu and van Staden 2003)	In Ayurvedic medicine for hypothyroidism (Ding and Staudinger 2005). As an emmenagogue, oral abortifacient (Lukhoba et al. 2006; Almeida and Lemonica 2000). In Somalia as an aphrodisiac (Lukhoba et al. 2006)	Throat and mouth infections, tonsillitis, gastrointestinal infections, genitourinary infections (e.g., syphilis in Central Africa), and eye and ear infections (Lukhoba et al. 2006). In Rwanda, Kenya, French Guiana, and Brazil to treat malaria (Lukhoba et al. 2006; Muhayimana et al. 1998; Vigneron et al. 2005). In Kenya for measles (Matu and van Staden 2003)

Table 3.2 Quantitative physicochemical values of *C. forskohlii* were collected from four geographical zones (Source Srivastava et al. 2002)

	Locations			
Parameters (percentage)	Tarikhet (Uttar Pradesh) Mean ± SD	Agrakhal (Uttar Pradesh) Mean ± SD	Vijaywada (Andhra Pradesh) Mean ± SD	Salem (Tamil Nadu) Mean ± SD
Loss on drying	10.70 ± 0.653	8.03 ± 0.157	9.74 ± 0.333	8.40 ± 0.200
Total ash	7.19 ± 0.373	7.82 ± 0.345	7.58 ± 0.372	8.61 ± 0.105
Acid-insol. ash	1.41 ± 0.226	0.96 ± 0.171	1.15 ± 0.085	2.26 ± 0.050
Tannin	0.76 ± 0.043	0.86 ± 0.082	0.46 ± 0.030	0.77 ± 0.012
Protein	12.68 ± 0.890	12.63 ± 0.325	7.97 ± 0.068	6.50 ± 0.170
Sugar	2.73 ± 0.388	2.30 ± 0.035	8.39 ± 0. 678	7.56 ± 0.909
Starch	12.00 ± 0.153	13.16 ± 0.870	20.07 ± 6.784	15.91 ± 4.354
Alcohol extractive (I.P.)	16.15 ± 0.691	16.93 ± 2.244	28.83 ± 1.033	27.42 ± 1.020
Water extractive (I.P.)	23.51 ± 0.385	27.40 ± 1.507	36.08 ± 1.319	32.00 ± 0.707
Successive extractives				
Hexane	4.88 ± 0.057	4.19 ± 0.205	4.27 ± 0.083	3.79 ± 0.222
Chloroform	5.55 ± 0.100	1.41 ± 0.130	2.23 ± 0.170	1.96 ± 0.115
Acetone	17.68 ± 0.119	4.44 ± 0.417	15.59 ± 0.395	17.35 ± 0.227
Alcohol	17.70 ± 0.064	6.34 ± 1.158	15.52 ± 0.282	16.52 ± 0.036
Water	17.13 ± 0.130	26.80 ± 1.045	23.50 ± 0.270	19.42 ± 0.393

3.3.5 Pain, Inflammation, and Musculoskeletal Problems

It is used to treat inflammation, abdominal and spasmodic pain, painful micturition (Lukhoba et al. 2006; Dubey et al. 1981), muscular, generalized pain, stiff neck, backache, bone dislocation, and rheumatism (Lukhoba et al. 2006).

3.3.6 Metabolic and Endocrine System

In Ayurvedic medicine, *C. forskohlii* is used to treat hypothyroidism (Ding and Staudinger 2005). It is also used as an emmenagogue and oral abortifacient (Lukhoba et al. 2006; Almeida and Lemonica 2000). In Somalia it is used as an aphrodisiac (Lukhoba et al. 2006).

3.3.7 Cardiovascular System

It is also used to treat Angina, hemorrhage, and hypertension (Lukhoba et al. 2006; Dubey et al. 1981).

3.3.8 Sensory

It is used to treat conjunctivitis in Congo and earache in Kenya (Lukhoba et al. 2006).

3.3.9 Infection

C. forskohlii is used to treat throat and mouth infections, tonsillitis, gastrointestinal infections, genitourinary infections (e.g., syphilis in Central Africa), and eye and ear infections (Lukhoba et al. 2006). In Rwanda, Kenya, French Guiana, and Brazil it is used to treat malaria (Lukhoba et al. 2006; Muhayimana et al. 1998; Vigneron et al. 2005). In Kenya it is used to treat measles (Matu and van Staden 2003).

3.4 Pharmacognosy

Srivastava et al. (2002) described the pharmacognostic characters of roots of *C. forskohlii*. The following account is from this paper.

3.4.1 Macroscopic Characters of Root

Roots tuberous with pale brown color, tapering with few secondary roots, light in weight, 1.5–3.5 cm in diameter, wrinkled diagonally, cut portion of the root tuber surface with yellowish white color, fracture short, characteristic pleasing aromatic odor with bitter to pungent taste (Srivastava et al. 2002).

3.4.2 Microscopic Characters

A transverse section of the root is irregular in outline, with epidermal cells not visible due to secondary growth. The outermost layer is made up of multilayered rectangular cork cells. Under the cork cells 1–2 layers of cambium are present, followed by parenchymatous secondary cortical region made up of rectangular cells and it consists of sclereids and calcium oxalate crystals (Fig. 3.1a, b). A continuous ring of vascular cambium is present below the cork region (Fig. 3.2a, b). The phloem is made up of companion cells, sieve tubes, and phloem parenchyma. Well-developed and thin-walled medullary rays are arranged radially in the vascular region. The cells of medullary rays are varying in size and are heterogenous (Fig. 3.3).

Medullary rays in transverse section of the young root showed very wide in size than the cells of older roots. The xylem is composed of barrelshaped, solitary, and porous vessels. These vessels are rich in starch and showed reticulate to spiral thickenings. They also consist of starch grains with 20–60 mm diameter, distinct hilum, and star-shaped cleft at the central position. Tailed vessels are also observed.

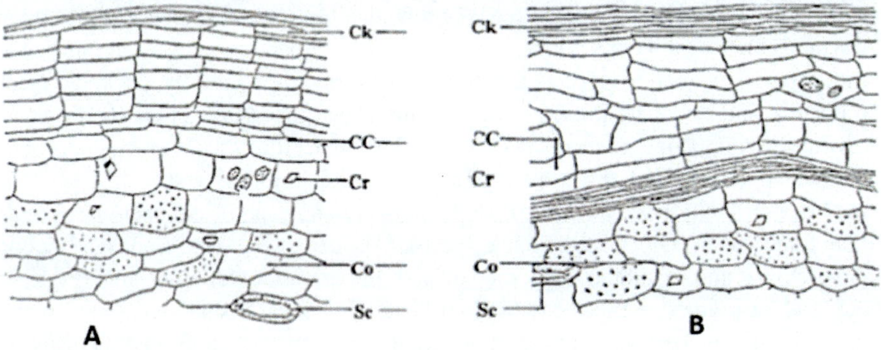

Fig. 3.1 C.S. of root showing sclereids in (**a**) young root (**b**) mature root (Source: Srivastava et al. 2002)

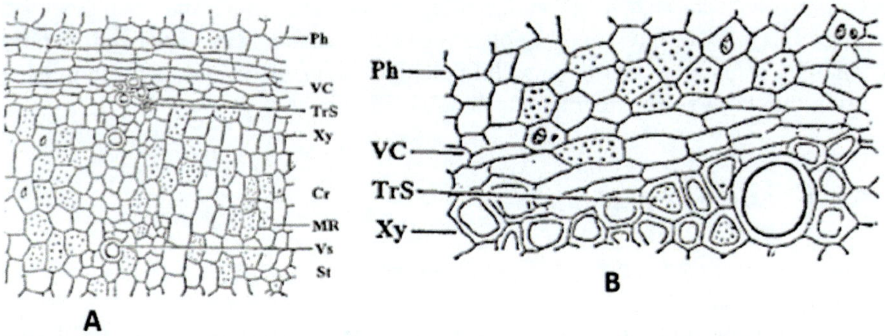

Fig. 3.2 C.S. of root showing vascular region in (**a**) young root (**b**) mature root (Source: Srivastava et al. 2002)

Fig. 3.3 Cellular details of LS of mature root (Source: Srivastava et al. 2002)

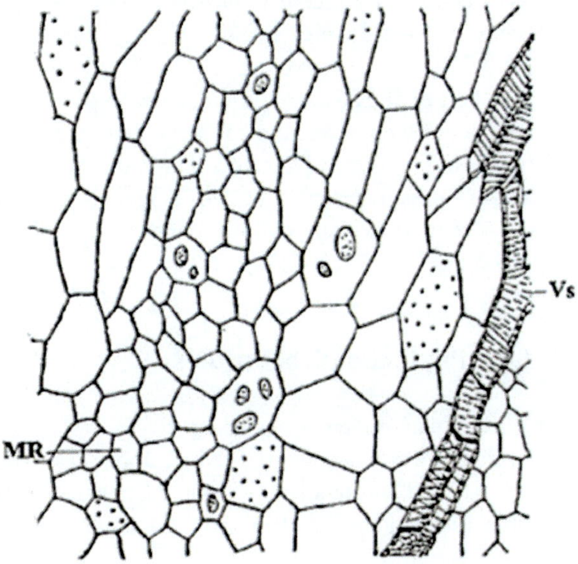

Tracheidal fibers and tracheids have bordered pits. Well-developed xylem parenchyma is present in young roots, whereas plenty of fibers are present in older roots. Pith portion is made up of parenchyma in young root, closely packed vessels, tracheids, and fibers are present in the mature roots (Srivastava et al. 2002).

Pharmacognostic studies of *C. forskohlii* leaf, root, and root tubers were conducted to localize the active principle of forskolin by using histochemical methods (Khatun et al. 2011a, b). Histochemical analysis of leaf sections revealed that forskolin was localized in leaf internal tissues such as palisade and spongy parenchyma and glandular trichomes of both upper and lower epidermis (Figs. 3.4, 3.5, and 3.6). The presence of forskolin in leaf tissues was visualized as brown color by staining with 10% vanillin-perchloric acid. It was further confirmed by using thin layer chromatography technique (Khatun et al. 2011a).

Histochemical analysis results on *C. forskohlii* root and root tubers revealed that forskolin was located in xylem, medullary rays, cortex, and cork cells. The presence of forskolin was stained as violet color in root (Fig. 3.7) and root tuber cells (Fig. 3.8a–c). The presence of forskolin in leaf tissues was visualized as brown color by staining with 10% vanillin-perchloric acid. It was further confirmed by using thin layer chromatography technique (Khatun et al. 2011b).

3.4.3 Study of Powder

The powdered root tuber is yellowish brown in color and pleasant aromatic odor with a bitter taste.

Microscopic examination of root powder shows oil globules, plenty of starch grains of circular, elliptical and ovoid shapes, sclereids, starch-filled vessels, tracheids, tailed vessels, parenchymatous patches, and calcium oxalate crystals (Fig. 3.9a–f). Root powder treated with 1 N NaOH + nitrocellulose saturated with amyl acetate turns greenish brown under UV light. Preliminary phytochemical analysis indicates the presence of alkaloids, flavonoids, and triterpenoids (Srivastava et al. 2002).

Tamboli et al. (2015) gave a detailed report on quality control of *C. forskohlii*, which includes physicochemical parameter determination, safety evaluation, microscopical evaluation, and chromatographic fingerprinting as well. Microscopic evaluation of transverse section of *Coleus* reveals that periderm, secondary phloem, and wide secondary xylem cylinder, which occupies a major portion of the root fragmentary.

3.4.4 Phytochemical Studies

The dried plant material was successively extracted with alcohol, acetone, hexane, chloroform, and water using Soxhlet apparatus. The percent yield of each extract was calculated and tabulated (Table 3.2). The quantity of total ash, water and acid soluble ash contents, and quantity of tannins, sugars, starch, and proteins were

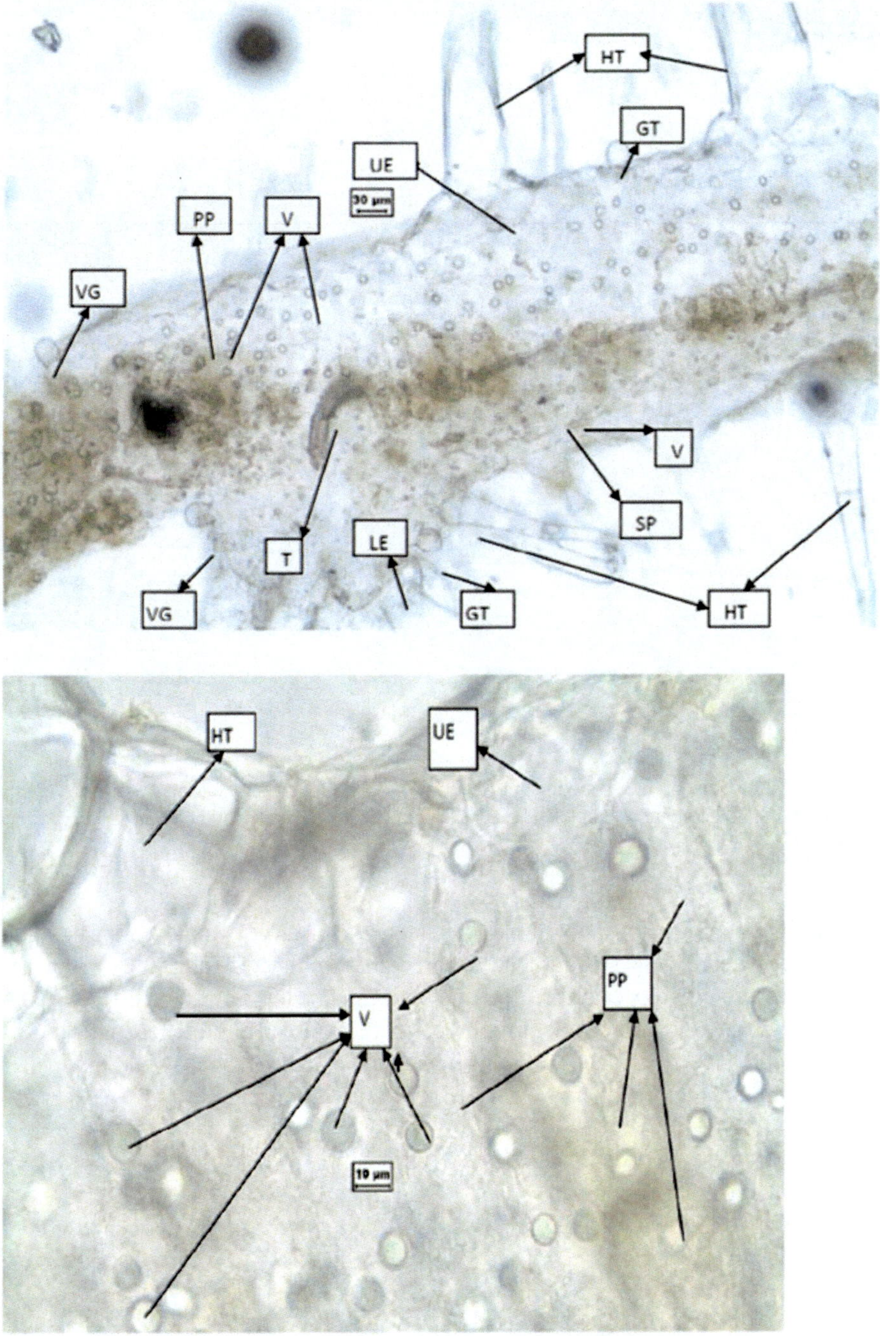

Fig. 3.4 *Coleus forskohlii*—Anatomy of leaf showing localization of forskolin in leaf internal tissues. *UE* upper epidermis, *V* violet-stained vesicles, *HT* hairy trichomes, *LE* lower epidermis, *PP* palisade parenchyma, *GT* glandular trichome, *VG* violet-stained gland, *T* trachea, *SP* spongy parenchyma (Source: Khatun et al. 2011a)

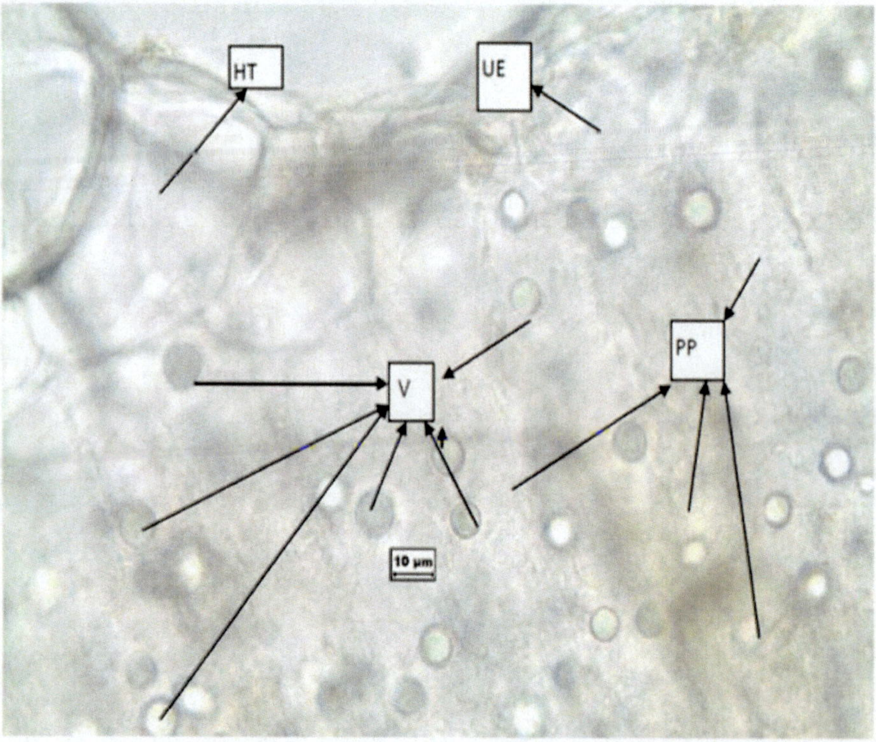

Fig. 3.5 Anatomy of leaf showing localization of forskolin in leaf internal tissues. *UE* upper epidermis, *V* violet-stained vesicles, *HT* hairy trichomes, *PP* palisade parenchyma (Source: Khatun et al. 2011a)

determined. Further, the amount of heavy metals such as Cu, Zn, Pb, Co, Mn, and Cr were determined with standard procedures and tabulated (Table 3.2). Thin layer chromatography was performed to develop a fingerprint profile for quality evaluation and standardization of the drug by using coleonol as a reference.

Disticraj and Jayaraman (2015) investigated the pharmacognosy and phytochemistry of *C. forskohlii.*

3.4.4.1 Powder Analysis of Whole Plant

Odor: Characteristic odor.

Color: Fresh powder pale green in color, dried powder grayish green in color.

Taste: Slightly acrid.

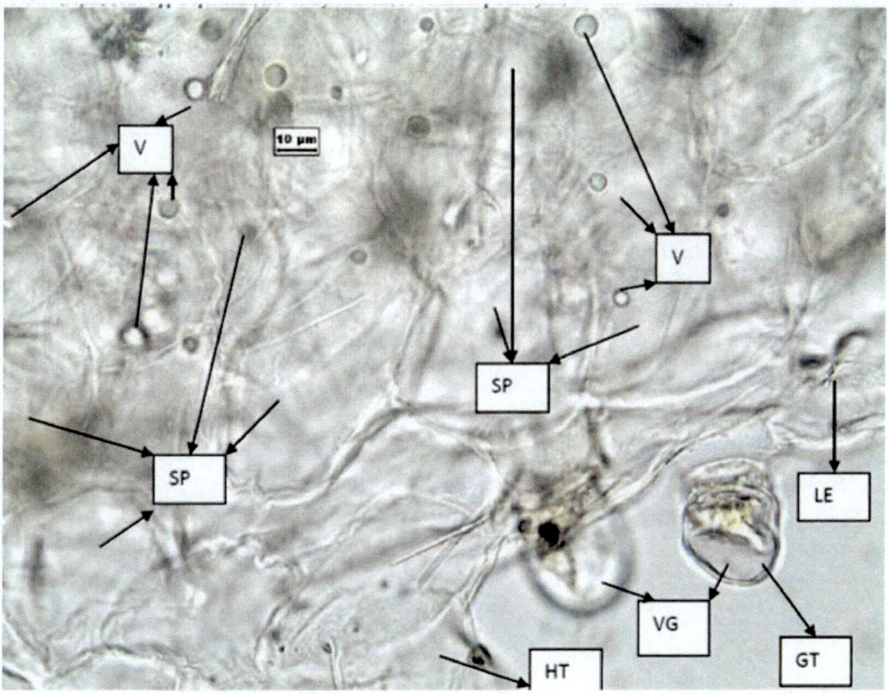

Fig. 3.6 Anatomy of leaf showing localization of forskolin in leaf internal tissues. *V* violet-stained vesicles, *HT* hairy trichomes, *LE* lower epidermis, *GT* glandular trichome, *VG* violet-stained gland, *SP* spongy parenchyma (Source: Khatun et al. 2011a)

3.4.4.2 Physicochemical Analysis of Whole Plant of *C. forskohlii* (Disticraj and Jayaraman 2015)

Average % of Physicochemical constants.

1. Total ash	1.23
2. Water soluble ash	2.22
3. Acid insoluble ash	2.04
4. Loss on drying	1.11
5. Water soluble extractive	0.78
6. Alcohol soluble extractive	1.67
7. Crude fiber content	1.03

Transverse section of fresh leaves of *C. forskohlii* showed the presence of very thin cuticles on epidermis, stomata on both lower and upper surfaces (Fig. 3.10). Glandular and nonglandular trichomes were observed on both epidermises. Palisade

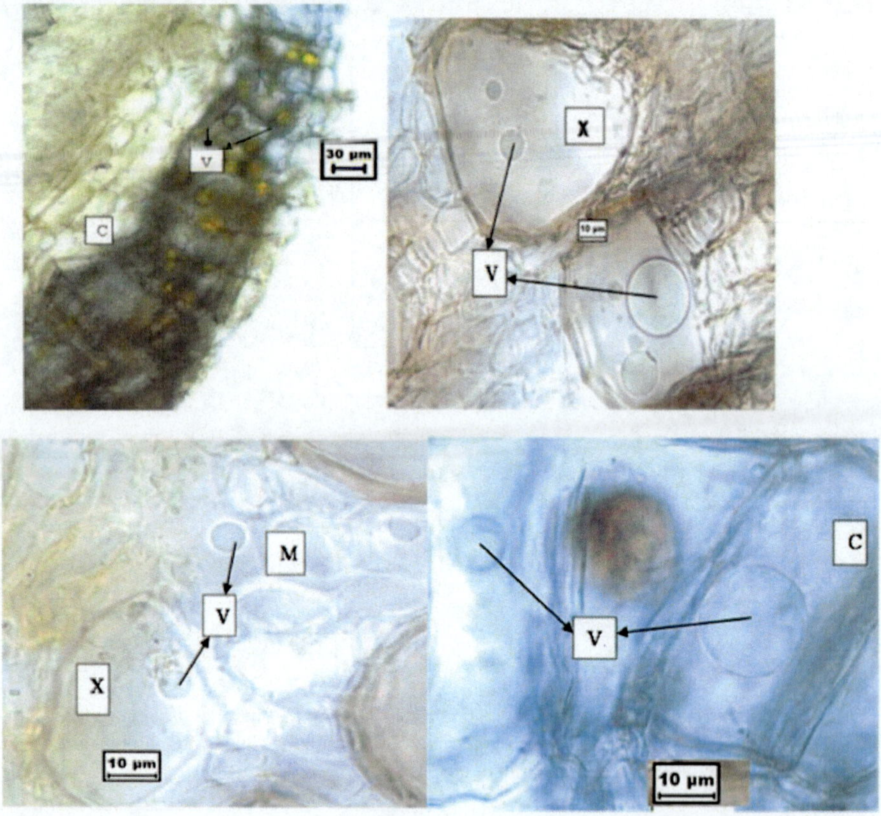

Fig. 3.7 Anatomy of root showing localization of forskolin in root internal tissues. *V* violet-stained vesicles, *C* cortex cells, *X* xylem cells, *M* medullary rays (Source: Khatun et al. 2011b)

tissue is present on both sides of the leaf. Multicellular heads with sessile glands, covering trichomes, spiral ducts, vessels, and starch grains were observed (Fig. 3.11a–g) (Ullah et al. 2013).

Study on the morphology and structures of trichomes of *C. forskohlii* of China origin, revealed that, four types of morphologically distinct trichomes were observed under a microscope (Huang et al. 2021). They are micro and macro types of glandular trichomes, peltate glandular trichomes, and nonglandular types of trichomes (Fig. 3.12a–d) observed on aerial parts of *C. forskohlii* growing in China (Huang et al. 2021).

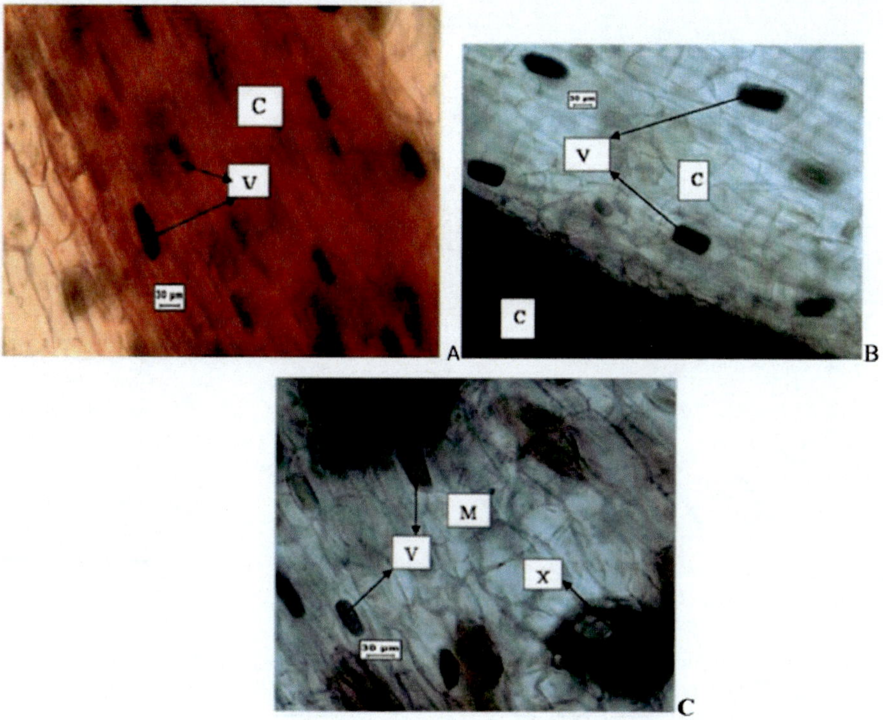

Fig. 3.8 (**a**) Cross section of root tuber showing localization of forskolin in root internal tissues. (**b**) Cross section of root tuber showing localization of forskolin in root internal tissues. (**c**) Cross section of root tuber showing localization of forskolin in root internal tissues. *V* violet-stained vesicles, *C* cork cells, *X* xylem cells, *M* medullary rays (Source: Khatun et al. 2011b)

3.5 Conclusion

The Pharmacognostic studies of various parts of *C. forskohlii* have been helpful in the correct identification of species if possible varietal differentiation based on the tissue differentiation and staining of major phytochemicals in the secretory cells. The distribution of xylem and phloem elements and other structural parts in the internal organs such as roots and stems for quality control. These studies also help in the mapping of high-yielding varieties based on anatomical features.

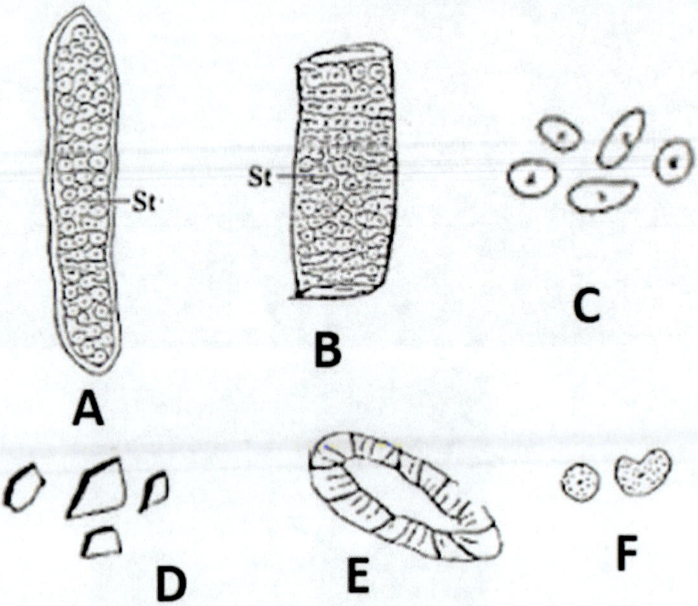

Fig. 3.9 Microscopic observations of powdered root. (**a**) Tracheid filled with starch, (**b**) Vessel filled with starch, (**c**) Starch grains, (**d**) Rhomboidal calcium oxalate crystals, (**e**) Sclereid, and (**f**) Oil globules (Source: Srivastava et al. 2002)

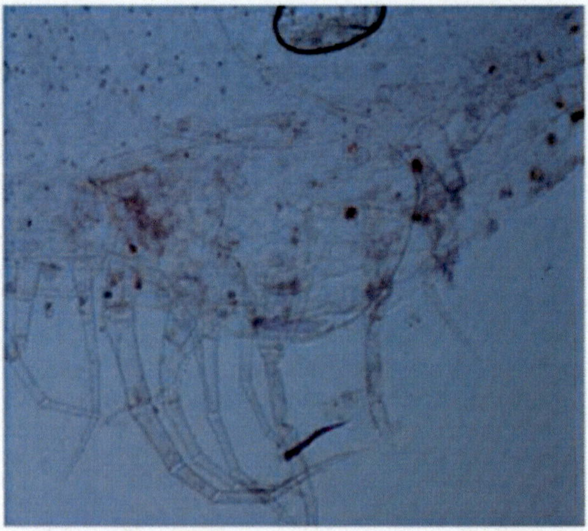

Fig. 3.10 Transverse section of lamina of the leaf (Source: Ullah et al. 2013)

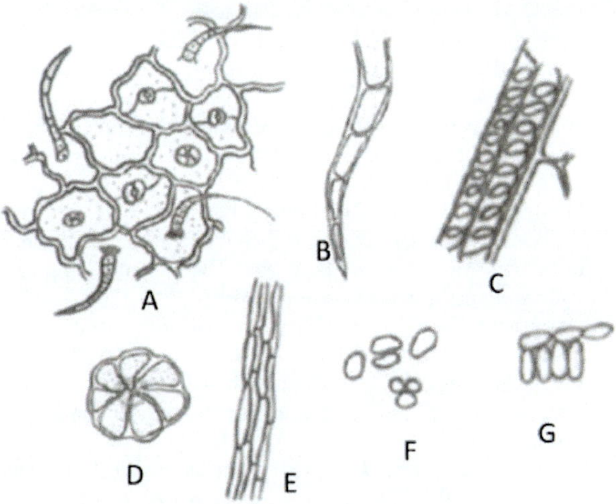

Fig. 3.11 Epidermal characters of *C. forskohlii* leaf (**a**) Epidermis with stomata, sessile glandular, and covering trichomes; (**b**) Covering trichomes; (**c**) Spiral ducts; (**d**) Multicellular head of sessile gland; (**e**) Vessels; (**f**) Starch grains; and (**g**) Palisade cells (Source: Ullah et al. 2013)

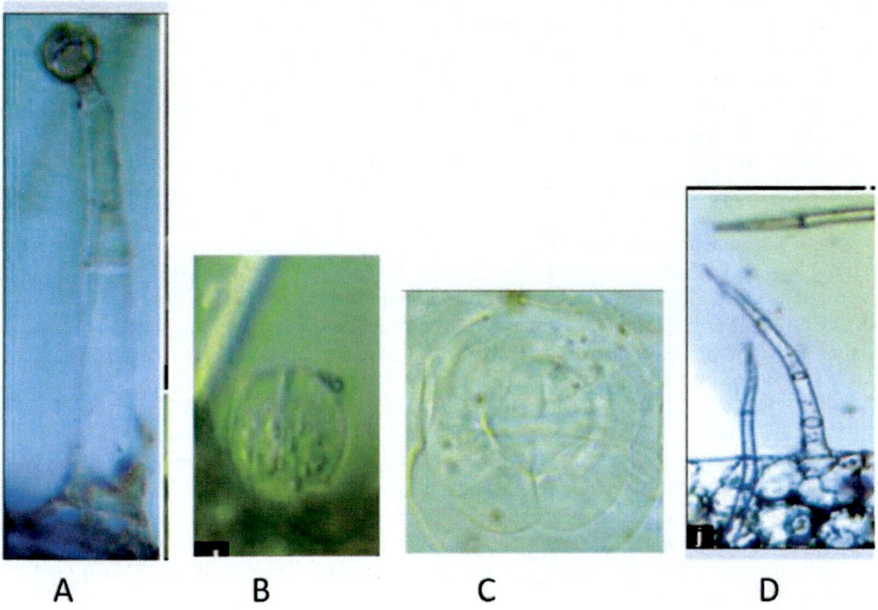

Fig. 3.12 Trichomes types were observed on *C. forskohlii*. (**a**) Glandular capitate macro type, (**b**) Glandular capitate micro type, (**c**) Peltate glandular trichomes, and (**d**) Multicellular nonglandular trichomes (Source: Huang et al. 2021)

References

Alabashi RH, Melzig MF (2010) *Plectranthus barbatus*: a review of phytochemistry, ethnobotanical uses and pharmacology- Part I. Planta Med 76:653–661

Almeida FC, Lemonica IP (2000) The toxic effects of *Coleus barbatus* on the different periods of pregnancy in rats. J Ethnopharmacol 73:53–60

Ammon HPT, Muller AB (1985) Forskolin: from an ayurvedic remedy to a modern agent. Planta Med 46:473–477

Camara CC, Nascimento NR, Macedo-Filho CL, Almeida FB, Fonteles MC (2003) Antispasmodic effect of the essential oil of *Plectranthus barbatus* and some major constituents on the Guinea-pig ileum. Planta Med 69:1080–1085

Chifundera K (2001) Contribution to the inventory of medicinal plants from the Bushi area, South Kivu Province, Democratic Republic of Congo. Fitoterapia 72:351–368

Chowdhary AR, Sharma ML (1998) GC-MS investigations on the essential oil from *Coleus forskohlii* Briq. Indian Perfumer 42:15–16

De Souza NJ, Shah V (1988) Forskolin—an adenylate cyclase activating drug from an Indian herb. Econ Med Plant Res 2:1–16

Ding X, Staudinger JL (2005) Induction of drug metabolism by forskolin: the role of the pregnane X receptor and the protein kinase A signal transduction pathway. J Pharmacol Exp Ther 312:849–856

Disticraj S, Jayaraman P (2015) Pharmacognostical and phytochemical analysis of *Asparagus racemosus* Willd, *Anisomeles malabarica* (L) R.Br., *Cleome monophylla* L. and *Coleus forskohlii* L. Int J Pharm Biol Sci 5(4):61–66

Dubey MP, Srimal RC, Nityanand S, Dhawan BN (1981) Pharmacological studies on coleonol, a hypotensive diterpene from *Coleus forskohlii*. J Ethnopharmacol 3:1–13

Fischman LA, Skorupa LA, Souccar C, Lapa A (1991) The water extract of *Coleus barbatus* Benth decreases gastric secretion in rats. Mem Inst Oswaldo Cruz 86(Suppl. 2):141–143

Githinji CW, Kokwaro JO (1993) Ethnomedical study of major species in the family Labiatae from Kenya. J Ethnopharmacol 39:197–203

Gupta S (2004) Nutracosmetics: a new growth segment. http://www.Happi.com/

Hamill FA, Apio S, Mubiru NK, Bukenya-Ziraba R, Mosango M, Maganyi OW, Soejarto DD (2003) Traditional herbal drugs of southern Uganda (II). Literature analysis and antimicrobial assays. J Ethnopharmacol 84:57–78

Huang X, You P, Chen Y, Zhang T, Gong L, Zou P et al (2021) Trichomes of *Coleus forskohlii*: morphology and volatile compounds. Pharmacogn Mag 17:558–564

Jin QD, He BH (1998) Minor constituents from *Coleus forskohlii*. Acta Bot Yunn 20:469–473

Johns T, Kokwaro JO, Kimani EK (1990) Herbal remedies of the Luo of Siya district, Kenya: quantitative criteria for consensus. Econ Bot 44:369–381

Kanne H, Burte NP, Prasanna V, Gujjula R (2015) Extraction and elemental analysis of *Coleus forskohlii* extract. Pharm Res 7:237–241

Kelecom A (1983) Isolation, structure determination, and absolute configuration of barbatusol, a new bioactive diterpene with a rearranged abietane skeleton from the labiate *Coleus barbatus*. Tetrahedron 39:3603–3608

Khatun S, Cakilcioglu U, Chatterjee NC (2011a) Pharmacognostic value of leaf anatomy and trichome morphology for identification of forskolin in a novel medicinal plant *Coleus forskohlii*. Biol Divers Conserv 4(2):165–171

Khatun S, Cakilcioglu U, Chatterjee NC (2011b) Phytochemical constituents vis-a-vis histochemical localization of forskolin in a medicinal plant *Coleus forskohlii* Briq. J Med Plants Res 5(5):711–718

Lukhoba CW, Simmonds MSJ, Paton AJ (2006) *Plectranthus*: a review of ethnobotanical uses. J Ethnopharmacol 103:1–24

Majeed M, Prakash L (2007) Fighting acne and more effective natural approaches to skin care. Cosmet Toilet:215–219

Matu EN, van Staden J (2003) Antibacterial and anti-inflammatory activities of some plants used for medicinal purposes in Kenya. J Ethnopharmacol 87:35–41

Misra LN, Tyagi BR, Ahmad A, Bahl JR (1994) Variability in the chemical composition of the essential oil of *Coleus forskohlii* genotypes. J Essent Oil Res 6:243–247

Muhayimana A, Chalchat JC, Garry RP (1998) Chemical composition of essential oils of some medicinal plants from Rwanda. J Essent Oil Res 10:251–259

Runyoro DK, Ngassapa OD, Matee MI, Joseph CC, Moshi MJ (2006) Medicinal plants used by Tanzanian traditional healers in the management of Candida infections. J Ethnopharmacol 106(2):158–165

Schultz C, Bossolani MP, Torres LMB, Lima-Landman MTR, Lapa AJ, Souccar C (2007) Inhibition of the gastric H+, K+-ATPase by plectrinone A, a diterpenoid isolated from *Plectranthus barbatus* Andrews. J Ethnopharmacol 111: 1–7

Srivastava SK, Chaubey M, Khatoon S, Rawat AKS, Mehrotra S (2002) Pharmacognostic evaluation of *Coleus forskohlii*. Pharm Biol 40(2):129–134

Tamboli ET, Chester K, Ahmad S (2015) Quality control aspects of herbs and botanicals in developing countries: *Coleus forskohlii* Briq a case study. J Pharm Bioallied Sci 7:254–259

Thorne Research (2006) Monograph on *Coleus forskohlii*. Altern Med Rev 11(1):47–51

Ullah S, Ibrar M, Barkatullah N, Rooh Ullah M (2013) Pharmacognostic, larvicidal and phytotoxic profile of *Coleus forskohlii* and *Rosmarinus officinalis*. J Pharmacogn Phytother 5(4):59–63

Valdes LJ, Mislankar SG, Paul AG (1987) *Coleus barbatus (C. forskohlii)* (Lamiaceae) and the potential new drug forskolin (Coleonol). Econ Bot 44:474–483

Vigneron M, Deparis X, Deharo E, Bourdy G (2005) Antimalarial remedies in French Guiana: a knowledge attitudes and practices study. J Ethnopharmacol 98:351–360

Yang QR, Wu HZ, Wang XM, Zou GA, Liu YW (2006) Three new diterpenoids from *Coleus forskohlii* Briq. J Asian Nat Prod Res 8:355–360

Chemistry of *Coleus forskohlii* Briq.

4

Abstract

Diterpenoid, forskolin is extracted from the roots of *Coleus forskohlii* Briq. (Syn. *C. barbatus* Benth. Lamiaceae). This species is the only source of forskolin. Many other phytochemicals have been isolated from leaves and roots. This chapter gives an account of diversified phytochemicals of terpenes, biosynthesis of forskolin, extraction, and separation of forskolin. Interestingly only a few compounds in the flavonoids were characterized. Various phytochemicals isolated from different parts of *C. forskohlii* have also been given. The biosynthetic process of forskolin has been discussed in addition to the scientific methods for its isolation and purification from stems and roots. The probable biosynthetic path of genkwanin is given. The biotechnological intervention and improvement methods of forskolin production in microbial models are also discussed.

Keywords

Coleus forskohlii · Forskolin · Coleonol · Barbutasin · Plectrin · Biogenesis · Phytochemicals · Diterpenes

4.1 Introduction

Alasbahi and Melzig (2010) reviewed in detail about the phytochemistry, medicinal uses as well as ethnobotanical uses along with the pharmacology of *Coleus forskohlii* Briq. (Syn. *C. barbatus* Benth. Lamiaceae) (Table 4.1). Several scientific studies and reviews were published on the phytochemistry and pharmacology of *C. forskohlii*. Kavitha et al. (2010) and Bhowal and Mehta (2017) reviewed the botanical, phytochemical, and pharmacological profiles and referred to as a nontoxic natural medicament when applied in recommended dose. In addition, Forskolin is regarded as safe even with LD_{50} value of 3100 mg/kg BW. It was reported to delay the fetal

T. Pullaiah, *Forskolin*, https://doi.org/10.1007/978-981-19-6521-0_4

Table 4.1 Diterpenoids isolated from *C. forskohlii* (Source: Alasbahi and Melzig 2010)

No. of the compound	Name of the compound	Part used	*C. forskohlii* location	References
1	(+)-Allylroyleanone (plectranthone J)	L	East Africa Kenya	Ruedi (1986)
2	Coleon S	L	China	Yao and Xu (2001); Yao et al. (2002)
3	Coleon O	L	East Africa—Kenya	Kubo et al. (1984)
4	Coleon T	L	China	Yao and Xu (2001); Yao et al. (2002)
5	Plectrin	L	East Africa—Kenya	Ruedi (1986); Kubo et al. (1984)
6	Barbatusin	L	Brazil	Zelnik et al. (1977); de Albuquerque et al. (2007); Wang et al. (1973)
7	3β-Hydroxy-3-deoxybarbatusin	L	Brazil	Zelnik et al. (1977)
8	Cyclobutatisin	L	Brazil	Zelnik et al. (1977); de Albuquerque et al. (2007); Wang et al. (1974)
9	7β-Acetyl-12-deacetoxycyclobutatusin	L	Brazil	de Albuquerque et al. (2007)
10	(16R)-Coleon E	L	East Africa—Kenya	Ruedi (1986); Ruedi and Eugster (1972)
11	Coleon F	L	East Africa—Kenya	Ruedi (1986); Ruedi and Eugster (1973)
12	(16R)-Plectrinon A	L	Brazil, East Africa—Kenya	Schultz et al. (2007); Ruedi (1986)
13	Plectrinon B	L	East Africa—Kenya	Ruedi (1986)
14	14-Deoxycoleon U	R	China	Xu et al. (2005)
15	Coleon C	WP	China	Liu et al. (2007)
16	6,7-Secoabietane diterpene I	S	Brazil	Kelecom et al. (1987)
17	6,7-Secoabietane diterpene II	S	Brazil	Kelecom et al. (1987)

(continued)

Table 4.1 (continued)

No. of the compound	Name of the compound	Part used	*C. forskohlii* location	References
18	Cariocal	S	Brazil	Kelecom and dos Santos (1985)
19	Abietatriene (dehydroabietane)	R	India	Mathela et al. (1986)
20	Demethylcryptojaponol (11-hydroxysugiol)	R	China	Xu et al. (2005)
21	Ferruginol	S	Brazil	Kelecom (1983a, b)
22	Sugiol	WP	China	Li et al. (2006)
23	20-Deoxocarnosol	S	Brazil	Kelecom et al. (1986)
24	6β-Hydroxycarnosol	S	Brazil	
25	Barbatusol	S	Brazil	Kelecom (1983a, b)
8,13-Epoxylabd-14-en-11-one-diterpenoids				
26	Forskolin (7β-acetoxy-1α,6β,9α-trihydroxy-8,13-epoxy-labd-14-en-11-one; coleonol; colforsin; 1-deacetyl forskolin B, 6-deacetylforskolin J)	R,R	India, China	Bhat et al. (1977a, b); Shah et al. (1980); Gabetta et al. (1989); Liu et al. (1992); Zhang et al. (2005)
27	9-Deoxyforskolin (7β-acetoxy-1α,6β-dihydroxy-8,13-epoxy-labd- 14-en-11-one)	R	India	Gabetta et al. (1989); Bhat et al. (1983)
28	1,9-Dideoxyforskolin (7β-acetoxy-6β-hydroxy-8,13-epoxy-labd-14-en-11- one)	R	India	Bhat et al. (1977a, b); Gabetta et al. (1989)
29	1,9-Dideoxy-7-deacetylforskolin (6β,7β-dihydroxy-8,13-epoxy-labd-14-en- 11-one)	R	India	Bhat et al. (1977a, b); Gabetta et al. (1989)
30	Deacetyl-1-deoxyforskolin (6β, 7β, 9α-trihydroxy-8,13-epoxy-labd-14-en-l lone)	R	India	Gabetta et al. (1989)
31	6-Acetyl-1-deoxyforskolin	WP	China	Xu et al. (2006)
32	6-Acetyl-1,9-dideoxyforskolin	WP	China	Xu et al. (2006)
33	1,6-Di-O-acetylforskolin (1α,6β,7β triacetoxy-9α-hydroxy-8,13-epoxy-labd- 14-en-11-one; forskolin A; 1,7-diaceylisoforskolin)	R, WP	China	Jin and He (1998); Wu et al. (2005); Jin et al. (1990)

(continued)

Table 4.1 (continued)

No. of the compound	Name of the compound	Part used	*C. forskohlii* location	References
34	1-Acetylforskolin (1α,7β-diacetoxy-6β, 9α-dihydroxy 8,13-epoxy-labd-14-en-11-one; forskolin B)	R, WP	China	Jin and He (1998); Wu et al. (2005); Jin et al. (1990)
35	Isoforskolin (6β-acetoxy-1α,7β,9α-trihydroxy-8,13-epoxy-labd-14-en-11-one; coleonol B; forskolin C; 1-deacetylforskolin I)	R, R, L, WP	India, China	Jin and He (1998); Bhat et al. (1977a, b); Zhang et al. (2005); Wu et al. (2005); Jin et al. (1990); Carpy et al. (1991); Prakash et al. (1988); Xu and Kong (2004); Yang et al. (2006); Feng et al. (2016)
36	1,9-Dideoxycoleonol B (7β-hydroxy-6β-acetoxy-8,13-epoxy-labd-14-en-11-one)	R	India	Roy et al. (1993)
37	7-Deacetylforskolin (1α,6β,7β,9α-tetrahydroxy-8,13-epoxy-labd-14-en-11- one; deacetylforskolin; 6-deacetylisoforskolin; forskolin D)	R, R, WP	India, China	Jin and He (1998); Bhat et al. (1977a, b); Gabetta et al. (1989); Wu et al. (2005); Jin et al. (1990)
38	Forskolin E (lα,7β-diacetoxy-6β-hydroxy-8,13-epoxy-labd-14-en-l1-one; 9-dehydroxyforskolin B)	R, WP	China	Jin and He (1998); Xu et al. (2006)
39	Forskolin F (7β-acetoxy-6β,9α-dihydroxy-8,13-epoxy-labd-14-en-11-one; 1-deoxyforskolin; 1-deacetoxyforskolin B; coleonol D)	R, R, WP	India, China	Jin and He (1998); Gabetta et al. (1989); Prakash et al. (1988); Xu et al. (2006); Khandelwal et al. (1989)
40	Forskolin G (1α-hydroxy-6β,7β-diacetoxy-8,13-epoxy-labd-14-en-11-one; l-deacetyl-9-dehydroxyforskolin A; l-deacetyl-6-acetylforskolin E)	R, WP	China	Xu and Kong (2004); Yang et al. (2006); Xu et al. (2006); Shen et al. (2002); Shan et al. (2006)
41	Forskolin H (1α,6β-diacetoxy-8,13-epoxy-labd-14-en-11-one; 7-deacetoxy-9-dehydroxyforskolin A; plectromatin C)	R, WP	China	Xu and Kong (2004); Yang et al. (2006); Xu et al. (2006); Shen et al. (2002)

(continued)

Table 4.1 (continued)

No. of the compound	Name of the compound	Part used	*C. forskohlii* location	References
42	Forskolin I (1α,6β-diacetoxy-7β,9α-dihydroxy-8,13-epoxy-labd-14-en-11-one; 7-deacetylforskolin A; 1-acetylforskolin C)	R, WP	China	Xu and Kong (2004); Yang et al. (2006); Shen and Xu (2005); Yang et al. (2007)
43	Forskolin J (1α,9α-dihydroxy-6β,7β-diacetoxy-8,13-epoxy-labd-14-en-11-one; 6-O-acetylforskolin;1-deacetylforskolin A; 7-acetylforskolin C)	R	China	Xu and Kong (2004); Shen and Xu (2005); Yang et al. (2007)
44	1,6-Diacetoxy-9-deoxyforskolin (1α,6β, 7β-triacetoxy-8,13-epoxy-labd-14- en-11-one; forskolin K; 9-dehydoxyfoskolin A)	R, WP	China	Li et al. (2006); Xu and Kong (2004); Yang et al. (2007)
45	6β-Hydroxy-8,13-epoxy-labd-14-en-11-one (forskolin L)	R, R	China, India	Gabetta et al. (1989); Xu and Kong (2004); Yang et al. (2007)
46	Coleosol (6β,9β-dihydroxy-8,13-epoxy-labd-14-en-11-one; 6β,9β-dihydroxy-11-oxomanoyloxide)	R	India	Prakash et al. (1988); Jauhari et al. (1978)
47	1-Acetoxy coleosol (1α-acetoxy-6β,9α-dihydroxy-8,13-epoxy-labd-14-en-11-one)	R	India	Roy et al. (1993)
48	Coleol (9α-hydroxy-8,13-epoxy-labd-14-en-11-one)	R	India	Gabetta et al. (1989); Prakash et al. (1988); Singh et al. (1984); Katti et al. (1979)
49	11-Oxomanoyloxide (8,13-epoxy-labd-14-en-11-one)	R	India	Gabetta et al. (1989)
50	Coleonol E (7α-acetoxy-6β-hydroxy-8,13-epoxy-labd-14-en-11-one)	R	India	Painuly et al. (1979)
51	Coleonol F (6β-acetoxy-7α,9α-dihydroxy-8,13-epoxy-labd-14-en-11-one)	R	India	Painuly et al. (1979)
52	Deoxycoleonol (7α-acetoxy-1α,6β-dihydroxy-8,13-epoxy-labd-14-en-11-one)	R	India	Tandon et al. (1978)
8,13-Epoxy-labd-diterpenoids with some deviations				
53	3-Hydroxyforskolin	WP	China	Shan and Kong (2006)

(continued)

Table 4.1 (continued)

No. of the compound	Name of the compound	Part used	*C. forskohlii* location	References
54	3-Hydroxyisoforskolin	WP	China	Shan and Kong (2006)
55	13-Epi-9-deoxycoleonol (13-epi-9-deoxyforskolin; 7β-acetoxy-1α,6β-dihydroxy-8,13-epoxy-labd-14-en-11-one)	R	India	Tandon et al. (1992)
56	Coleonol C (6β-acetoxy-1α,7α,9α-trihydroxy-8,13-β-epoxy-labd-14-en-11-one)	R	India	Tandon et al. (1978)
57	Coleonone (8,13-epoxy-labd-14-en-12-one)	R	India	Singh et al. (1984); Katti et al. (1979)
58	Manoyl oxide (8,13-epoxy-labd-14-ene)	R	India	Mathela et al. (1986)
Miscellaneous labdane diterpenoids				
59	13-Epi-sclareol	R	India	Sashidhara et al. (2007)
60	Forskoditerpene A (5β,9β,10α,12β-9,12-cyclo-7,13E-labdadien-15-oic acid)	WP	China	Shan et al. (2008)
61	12-Hydroxy-8,13E-labdadien-15-oic acid	WP	China	Xu and Kong (2006)
62	Coleolic acid (11-ol,13-Me, 8(9),13(14)Z-labdadien-15-oic acid)	WP	China	Liu et al. (2007)
63	Coleonic acid (11-one,13-Me, 8(9),13(14)Z-labdadien-15-oic acid)	WP	China	Liu et al. (2007)
8,13-Epoxy-labd-14-en-11-one-diterpene glycosides				
64	Forskoditerpenoside A (6β-acetoxy-7β,9α-dihydroxy-8,13-epoxy-labd-14-en-11-one-1α-O-β-D-glucopyranoside)	WP	China	Shan et al. (2007)
65	Forskoditerpenoside B (6β,7β-diacetoxy-9α-hydroxy-8,13-epoxy-labd-14-en-11-one-1α-O-β-D-glucopyranoside)	WP	China	Shan et al. (2007)
66	Forskoditerpenoside C (6β-acetoxy-7β-hydroxy-8,13-epoxy-labd-14-en-11-one-1α -O-β-D-glucopyranoside)	WP	China	Shan et al. (2008)
67	Forskoditerpenoside D (6β, 7β-diacetoxy- 8,13-epoxy-labd-14-en-11-one-1α -O-β-D-glucopyranoside)	WP	China	Shan et al. (2008)

(continued)

Table 4.1 (continued)

No. of the compound	Name of the compound	Part used	*C. forskohlii* location	References
68	Forskoditerpenoside E (6β-acetoxy-8,13-epoxy-labd-14-en-11-one-1α-O-β-Dglucopyranoside)	WP	China	Shan et al. (2008)
Other assorted compounds				
69	Thymoquinol-2-*O*-*β*-glucopyranoside	WP	Saudi Arabia	Shaker et al. (2022)
70	Syringic acid	WP	Saudi Arabia	Shaker et al. (2022)
71	Methyl 3,4,5-trihydroxybenzoate,	WP	Saudi Arabia	Shaker et al. (2022)
72	Luteolin	WP	Saudi Arabia	Shaker et al. (2022)
73	Cytochalasin B	–	China	Zhang and Kong (2009)
74	N-benzoyl-L-phenylalaninol	–	China	Zhang and Kong (2009)
75	3,6-Dibenzyl-2,5-dioxopiperazine	–	China	Zhang and Kong (2009)
76	2-Furoic acid	–	China	Zhang and Kong (2009)
77	Vanillic acid	–	China	Zhang and Kong (2009)
78	Loliolide	–	China	Zhang and Kong (2009)
79	Chamaecydin	–	China	Wang et al. (2009)
80	6 α-hydroxydemethylcryptojaponol	–	China	Wang et al. (2009)
81	α-Cedrene	–	China	Wang et al. (2009)
82	Lupeol	–	China	Huang et al. (2011)
83	Uvalo	–	China	Huang et al. (2011)
84	Dehydroabietane	AP	Saudi Arabia	Mothana et al. (2014)
85	5,6-Didehydro-7-hydroxy-taxodone	AP	Saudi Arabia	Mothana et al. (2014)
86	Taxodione	AP	Saudi Arabia	Mothana et al. (2014)
87	6α,11,12,-Trihydroxy-7β,20-epoxy-8,11,13-abietatriene	AP	Saudi Arabia	Mothana et al. (2014)
88	Barbaterpene (2'R-hydroxydocosanoylursa-12-en-3β-ol)	–	Saudi Arabia	Amina et al. (2018)

(continued)

Table 4.1 (continued)

No. of the compound	Name of the compound	Part used	*C. forskohlii* location	References
89	Barbatustero	–	Saudi Arabia	Amina et al. (2018)
90	1,2,3,4,6-Penta-O-galloyl-{beta}-D-glucose (PGG)	–	Brazil	dos Santos et al. (2012)
91	Plectrabarbene	–	–	Musayeib et al. (2020)

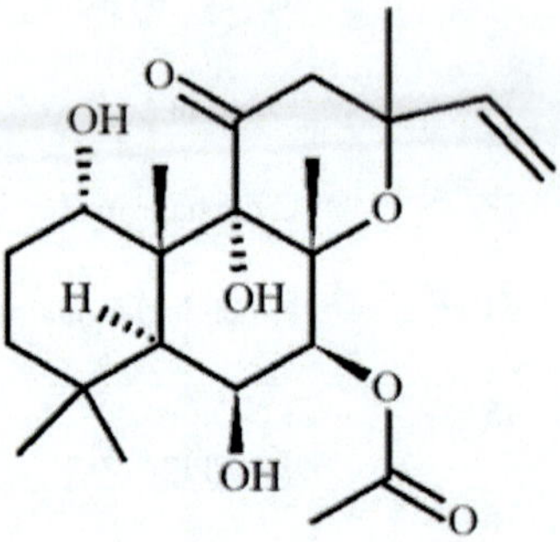

Fig. 4.1 Structure of forskolin

development and anti-implantation effect in pregnant rats at the dose of 880 mg/kg BW may be applied as an abortifacient natural drug (Almeida and Lemonica 2000).

Forskolin ($C_{22}H_{34}O_7$; Fig. 4.1) is a 7 β-acetoxy 8, 13- epoxy-1α, 6 β, and 9α-trihydroxy labd-14-en-11-one. The researchers of CDRI (Central Drug Research Institute), Lucknow, India between 1974 and1986, isolated and characterized 11-oxo-manoyl oxide, a diterpenoid from the roots of *C. forskohlii*. The effective and pharmacologically active diterpene was characterized and named "Coleonol". Thereafter de Souza, "Hoechst India" Limited, Mumbai called this compound as forskolin (Bhat et al. 1977a, b; Fig. 4.1) with similar pharmaceutical activities as Coleonol. The location of 7-acetoxy group made the difference between coleonol and forskolin. Forskolin is a semi-white solid crystal with a melting point of 228–230 °C and UV absorption maxima peaks such as 210 and 305 nm were reported. Tandon et al. (1979) reported variations in the quantity of Forskolin and other compounds using chromatographic analysis obtained from the samples from different countries such as Africa, India, and Brazil which has assigned for the climatic conditions.

An anti-inflammatory flavone, genkwanin (7-O-methylapigenin) was reported (Alasbahi and Melzig 2010). The flavonoids such as chrysosplenetin and 4,7-dimethoxy-5,6-dihydroxyflavone were also isolated from this plant (Gabetta et al. 1989). The minor components namely rosmarinic acid, scutellatrin 4′-methyl ether 7-O-glucuronide obtained from the polar extracts with a significant inhibition effect on the activity of acetylcholinesterase (Falé et al. 2009).

The altitudinal difference in the quantity of forskolin in the populations of Gopeswar with 1488 m MSL (mean sea level) to the Pipalkoti plants (Uttarakhand of India) with 1339 m MSL using HPLC studies. The Pipalkoti plants showed a high amount of forskolin (Rana et al. 2021). Isoforskolin, an analogue of forskolin, has been obtained by Feng et al. (2016) from Yunnan of China. Researchers found that the Yunnan native plant *Coleus foskohlii* contains rich isoforskolin but not foskolin.

4.2 Biogenesis

4.2.1 Forskolin

The biosynthesis of forskolin (Figs. 4.1 and 4.2) is noticed as acetate—mevalonate pathway. The compound namely, 8, 13-epoxy-labd-14-en-11-one is the initially produced labdane diterpenoid (mono oxygenated) and simultaneous inclusion of oxygen produces 1,9-dideoxy forskolin, 9-deoxyforskolin, and forskolin. However, a few more diversified terpenoids were synthesized besides Forskolin as the end product in the sequence (Akhila et al. 1990).

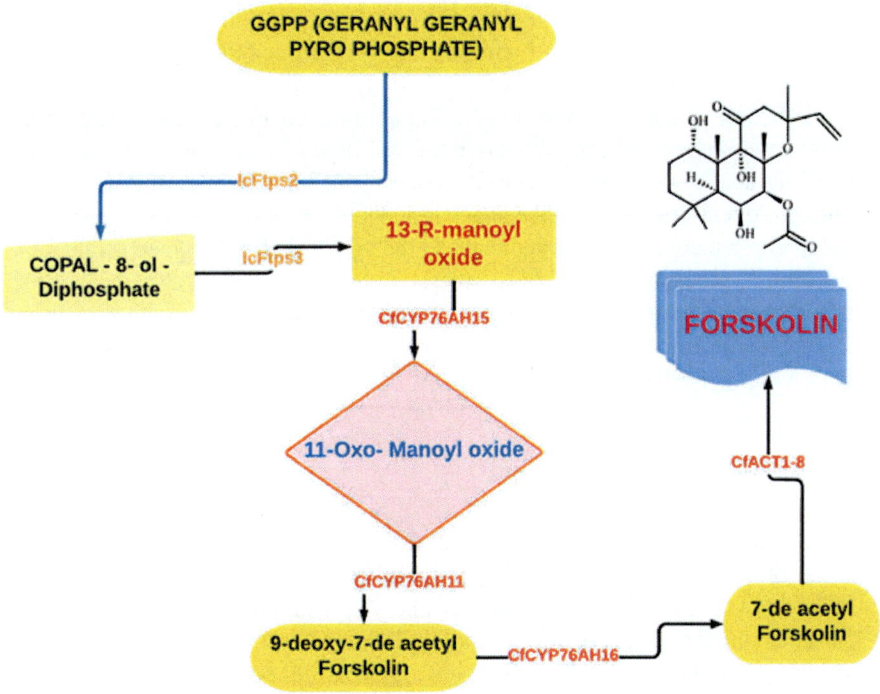

Fig. 4.2 Representative path of biosynthesis of forskolin

Isopentenyl diphosphate (IPP, a C-5 membered compound), is a metabolic precursor, from this two special pathways were initiated in the cytoplasm namely IPP and the mevalonate (MVA) paths besides an alternative pathway in the plastids namely 2C-methyl-D-erythritol 4 phosphate (MEP) pathway also being reported (Rohmer et al. 1993, 1996; Eisenreich et al. 1998). However, the IPP is biosynthetic initial metabolite to Forskolin was produced through non-mevalonate path. Here, Geranyl geranyl diphosphate (GGPP) synthase was identified as a crucial and key enzyme in the production of Forskolin, which is being synthesized in the leaves and transported to stems and roots (Engprasert et al. 2004). GGPP synthase (prenyl-transferase) catalyzes the reaction among allylic diphosphate condensing with the three molecules of IPP during production of GGPP. Wang and Ohnuma (1999) reported that Geranyl geranyl diphosphate is a linear molecule that is considered a precursor for the biosynthesis of diterpenes.

Although total synthesis of forskolin has been reported (Hashimoto et al. 1988), it is uneconomic due to structural complexity of the molecule. It is reported as not much effective than the naturally obtained forskolin, moreover *C. forskohlii* is reported as the only botanical source. Shah et al. (1980) studied that the forskolin was exclusively reported from *C. forskohlii* and it is not detected in its allied *Coleus* species viz., *C. amboinicus*, *C. blumei*, *C. canisus*, *C. malabaricus*, *C. parviflorus*, and *C. spicatus*. Forskolin was not detected in six taxonomical allied species in the Lamiaceae such as *Plectranthus* species namely, *P. coesta*, *P. incanus*, *P. melissoides*, *P. mollis*, *P. rugosus*, and *P. stocksii*. Further, analysis carried out on one hundred samples of *Coleus*, *Orthosiphon* and *Plectranthus* species of Ocimoideae (the sub family of Lamiaceae) also confirmed the absence of Forskolin in all the tested samples from Japan.

The structural complexity of Forskolin and its extraction as well as difficulty in the process of chemical synthesis are major problems in its commercial production not being able to meet the industrial needs. However, the scientific advancement boosted rapid improvement in the analysis and interpretation of biosynthetic paths in diterpene production, a novel approach to the synthesis for the green production of Forskolin. The heterologous biosynthesis was initiated by Yuan et al. (2020), through understanding its structure, activity and its biosynthesis through bio-engineering models leading to construction of new strains of bacteria for the high yielding of Forskolin.

4.2.2 Molecular Approach

Multiple approaches from different scholars lead to the identification of precursors for the biosynthesis of Forskolin from the *C. forskohlii* roots cultures (Bhat et al. 1977a, b; Asada et al. 2012). Zerbe et al. (2013) described its pathway through 454 and Illumina sequencing technology. Pateraki et al. (2014) reported the forskolin from the cork tissue of roots of *C. forskohlii* based on the histochemical tests and identified specialized cells possessing specific structures with oily organelles. The localization of forskolin confirmed a diterpene simplest backbone such as "(13R) manoyl oxide" in the specialized oily organelles. The "(13R) manoyl oxide" was reported to be synthesized in two specialized reactions favoured by two distinct class of enzyme belonging to Diterpene synthases.

Pateraki et al. (2014) studied the transcripts data and reported six functional "diterpene synthases" (diTPS) such as *Cf*TPS1 (KF444506), *Cf*TPS2 (KF444507), *Cf*TPS3 (KF444508), *Cf*TPS4 (KF444509), *Cf*TPS14 (AGN70881.1), and *Cf*TPS15 (KF4710011). These genes (diTPS) were heterologously translated in certain models such as *Escherichia coli* and *Nicotiana benthamiana* transient expressions. The chemical studies reported that the genes like *Cf*TPS1 and *Cf*TPS2 were functionally different from "type II diTPS" except the *Cf*TPS15 (yet to be reported and confirmed from the authors' firsthand observations). An enzymatic study showed that *Cf*TPS2 synthesized an intermediate compound "copal-8-ol diphosphate" in combination with *Cf*TPS3 resulting in the production of 13R-MO, a stereospecific metabolite. The combination of *Cf*TPS2 and *Cf*TPS4 was also reported for the production of 13R-MO and its epimer 13S-MO, whereas the other combinations failed to produce "manoyl oxide." The findings *Cf*TPS2 and *Cf*TPS3 led foundation for approaching forskolin biosynthetic pathway studies. Subsequently, cytochrome P450 enzymes (P450s) and acetyltransferases were reported and analyzed using metabolomics studies and other informatics tools of molecular modelling (Pateraki et al. 2017). The sequencing of RNA and analysis of *C. forskohlii* root tissue resulted in the discovery of genes like 263,652 cDNA. These studies noticed that 29 cytochrome P450 candidate genes were screened and found relative expression levels in the root tissue (cork). Seven genes were reported to belong to the CYP76AH subfamily, such as *Cf*CYP76AH8 (KT382348), *Cf*CYP76AH9 (KT382347), *Cf*CYP76AH10 (KT382346), *Cf*CYP76AH11 (KT382349), *Cf*CYP76AH15 (KT382358), *Cf*CYP76AH16 (KT382359), and *Cf*CYP76AH17 (KT382360). Two acetyltransferase (ACT) genes were also reported, like *Cf*ACT1–6 (KT382361) and *Cf*ACT1–8 (KT382363). Among these genes, five were analyzed and sequenced the full-lengths (*Cf*CYP76AH8, *Cf*CYP76AH9, *Cf*CYP76AH10, *Cf*CYP76AH11, and *Cf*CYP76AH17). However, the two genes like *Cf*CYP76AH15 and *Cf*CYP76AH16 were studied partially for its cDNAs.

The cytochrome P450s and acetyltransferase genes in *C. forskohlii* were functionally characterized using transient expression studies in *Nicotiana benthamiana*, and the intermediate metabolites were identified. The genes like *Cf*CYP76AH15, *Cf*CYP76AH8, and *Cf*CYP76AH17 can catalyze the reaction from 13R-MO to produce 11-oxo-manoyl oxide and found that the gene like "*Cf*CYP76AH15" as a highly efficient. The gene namely *Cf*CYP76AH8 and *Cf*CYP76AH17 also found effective in monooxidizing reaction at the C-1 site. The *Cf*CYP76AH11 favored the conversion of 13R-MO into "9-deoxy-7-deacetylforskolin." Whereas *Cf*CYP76AH16 catalyzed the transformation of 13R-MO into "9-hydroxymanoyl oxide" and *Cf*CYP76AH15 coupled with *Cf*CYP76AH11 and *Cf*CYP76AH16 catalyzed the transformation of 13R-MO into 7-deacetylforskolin. The enzyme

*Cf*CYP76AH15 catalyzed the acetylation of 7-deacylforskolin to form forskolin. Two ACT enzymes such as *Cf*ACT1–6 and *Cf*ACT1–8 catalyze the acetylation of 7-deacylforskolin. *Cf*ACT1–8 exhibited high specificity and activity, and effectively converted 7-deacylforskolin into forskolin (Ju et al. 2021).

4.2.2.1 Bioengineering Process for Biosynthesis of Forskolin

The utilization of microorganisms in the biosynthesis of commercially valuable and interesting natural products has been a regular practice. Ju et al. (2021) discussed the construction of cell factories to produce forskolin via synthetic biological methods. Expression of specific enzymes in the production of forskolin in various microbial models such as *Escherichia coli*, *Synechocystis* sp. PCC 6803 (Cyanobacteria) and *Saccharomyces cerevisiae* (yeast) have been utilized as tools for forskolin biosynthesis. The studies on the 13R-MO metabolic pathway in reconstructed strain *Synechocystis* sp. PCC 6803 has been reported for its optimized expression of the MEP pathway and through exploring the optimal induction conditions favored production of precursor compounds like 13R-MO (menoyl oxide) with a maximum yield of 0.45 mg/g DCW (Englund et al. 2015). Sutardja et al. (2019) demonstrated the production of the diterpenoid, forskolin (cAMP booster) using Cyanobacterium, *Synechocystis* PCC 6803 model and achieved enhanced production of 13R-MO by 15 times through the overexpression of key enzymes in the MEP pathway, as it did with expression of the terpene synthases alone.

4.2.2.2 Genkwanin

The only non-glycosylated flavone namely, genkwanin (7-O-methyl-apigenin) detected from *C. forskohlii* which proved its anti-inflammatory effects (Gao et al. 2014). The study about the biosynthesis of genkwanin has been reported to involve Flavone synthase expression and production (Jeon et al. 2009). It was well studied about the production of certain secondary metabolites through the expression of geranylgeranyl diphosphate synthase (GGPPS), 1-deoxy-D-xylulose-5-phosphate reducto-isomerase (DXR), Diterpene synthases, Cytochrome-P450 monooxygenases (CYP450s), and Chalcone synthase (Engprasert et al. 2004, 2005; Zerbe et al. 2013; Awasthi et al. 2015, 2016a, b).

A detailed study on the gene expression of genkwanin explained the relation of mannitol to the production of anthocyanin and genkwanin. The study revealed an inverse relationship in the expression of anthocyanin in the presence of mannitol (Fig. 4.3). The genes such as flavone synthase (CYP93B) and flavonoid 3 monooxygenase (CYP706C) are involved in the expression and synthesis of certain flavonoids in the leaf tissues. The maximum expression of the two genes of cytochrome P450 gene cassette (*CfCYP93B* and *CfCYP706C*) in leaf tissues correlated with the flavonoid synthesis. Whereas mannitol enhanced the expression of *CfCYP93B* and simultaneous depression in *CfCYP706C*. The quantification studies indicated that genkwanin has a positive correlation with the enhanced anthocyanin which has been affected by the mannitol treatment. The molecular docking approach along with certain protein modelling studies was supporting the

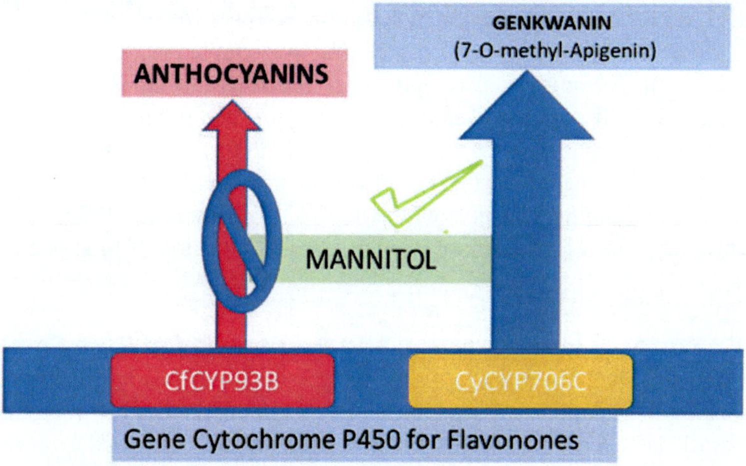

Fig. 4.3 Representation of effect of Mannitol (an elicitor) on gene expression of flavonoid (Genkwanin) biosynthesis in *C. forskohlii*

expression of *CfCYP93B* involvement in the transformation of naringenin to probable flavonoids namely genkwanin through apigenin. However, *CfCYP706C* inteferes in the mobilization of flavonoid precursors and channelizes toward synthesis of flavonols and anthocyanins.

Alignment, phylogenetic analysis, modelling, and molecular docking analysis of protein sequences suggested that *CfCYP93B* may be involved in the conversion of naringenin to flavones (possibly genkwanin through apigenin), while *CfCYP706C* may act on common precursors of flavonoid metabolism and channel the substrate toward production of flavonols or anthocynanins. Hence, it is understood that the production of Genkwanin and its accumulation depended on mannitol treatment which decreases the expression of *CfCYP706C* and even suppresses the production of other flavonoids in *C. forskohlii* (Awasthi et al. 2016c).

4.2.3 Extraction and Separation of Forskolin

The only natural source for forskolin is the tubers of *C. forskolii* and they are under commercial cultivation. After harvesting the tubers with 75–85% of moisture are stored after drying with less than 12% moisture. The plant tubers made into slices of 0.5 cm, are dried at 40 °C and packed in polythene-lined gunny bags for the maximum extraction of forskolin (Rajangam 2005). The standardization for the forskolin quantification studies using gas-liquid chromatography (GLC) by Inamdar et al. (1980) and later thin layer chromatography (TLC) and high-performance liquid chromatography (HPLC) were employed by Inamdar et al. (1984). Antibody-specific affinity chromatography was also employed for the quantification of forskolin in plant tissues during the developmental stages (Yanagihara et al. 1996). Later Demetzos et al. (2002) used the Nuclear magnetic resonance spectrum coupled with GC-MS for the quantification of forskolin. Another advanced technique namely RPLC (Reverse Phase Liquid Chromatography) coupled with a detector (photodiode array) at 210 nm was also reported as a sensitive technique for the detection from plants as well as market samples claiming for the forskolin as active principle (Schaneberg and Khan 2003). Later, Saleem et al. (2006) developed an economically feasible RPHPLC method using an adsorbent (activated charcoal) in the column which aids in obtaining high purity forskolin. Whereas Wu et al. (2007) developed a fingerprint approach for the quality control for the detection of forskolin.

4.2.4 Protocol for Separation and Preparation of Pure Forskolin

The method of isolation and separation of Forskolin from the raw material (stems) and extract has been discussed by Saleem et al. (2005). The dried, powdered stem material (up to 1000 μm) was extracted with chloroform continuously at 30 ± 5 °C through stirring for 2 h. The extract is separated and filtered and the filtrate has been concentrated at a reduced pressure which yields a semisolid substance. This residue loaded on silica gel column of 100 cm length ad 25 cm diameter with activated charcoal and toluene. The fractionation of residue performed on the solvent system used in the proportions of toluene and ethyl acetate in different ratios sequentially with the increasing polarity of the solvent combinations (Toluene: Ethyl acetate, 95: 5; 90:10; 90:20; 85:15; 80:20). The fractions of 80:20 ratio was carefully pooled and concentrated at reduced pressure (240–270 mbars) at 40 °C which results into a solid residue further crystallized in the ethyl acetate and n-hexane (1:15 ratio) and appears off-white. Saleem et al. (2005) confirmed this product as Forskolin through spectroscopic and analytical techniques and suggested a standardized protocol for isolation of forskolin. The yield of the forskolin obtained from dried stem powder has been reported as 100 mg/500 g or 0.0002%.

In another study by Saleem et al. (2006) reported isolation protocol from the roots using an alternate method to the column chromatography which may be a tedious process. In their study, 100 grams of dried root powder (up to 1000 μm) was continuously stirred at 60 °C with toluene (500 ml) and the extract was concentrated at reduced pressure at 40 °C to obtain 15 ml residue. The process was continued thrice and the extracts were pooled up and separated with n-hexane (150 ml), which results into a brownish semisolid precipitate of amorphous powder (1.6 grams). Later, a 10-ml syringe (7.5 cm length x 1.7 cm diameter) without a needle was taken and filled with the activated charcoal (6 grams) eluted with methanol using a Buchner flask. This protocol was repeated two times and connected with a vacuum motor. The elution with acetonitrile: methanol with the ratio of 1:1 for three times was followed by thin layer chromatography (TLC) to confirm the prominent spots of yellow color indicating the forskolin. The fractions of the similar spots on TLC were pooled and concentrated under reduced pressure which yields about 500 mg residue. This product was washed with n-hexane to remove nonpolar impurities for 2–3 instances and the remaining part was left for dryness. This part was dissolved in chloroform (5 ml) and again mixed with n-hexane to remove the coloring material until a pale-yellow product was obtained. Further, the Forskolin has been crystallized using the standardized protocols of Saleem et al. (2006) and examined in the reverse phase liquid chromatography or spectroscopic studies. This method has been encouraged as a simple, easy, and reliable for the isolation of Forskolin.

4.3 Phytocompounds Isolated from *C. forskohlii*

Roy et al. (1993) reported two new diterpenes namely, 1,9 dideoxy coleonol B and 1-acetoxy coleosol from the roots. Yao et al. (2002) characterized the two new diterpenoids, forskolin G and H from the chloroform fractions of the roots and elucidated the structures as 1α-hydroxy-6β,7β-diacetoxy-8,13-epoxylabd-14-ene-11-one and 1α,6β-diacetoxy-8,13-epoxylabd-14-ene-11-one. The new glycosides of labdane diterpenes such as Forskoditerpenoside A and B were isolated and characterized from the whole plant using ethanol extracts (Shan et al. 2007). Forskoditerpenoside A and B were demonstrated for the relaxative property in guinea pigs. Shan et al. (2008) reported three novel metabolites such as forskoditerpenoside C, D and E as minor components along with "Forskoditerpene A" from the ethanol extract and proved for the ralaxative effect in guinea pigs (Shan et al. 2008). Besides, Shan et al. (2008) also reported 8,13-epoxy-labd-14-en-11-one glycoside and Forskoditerpene A labdane derivatives from the *C. forskohlii*.

Shen and Xu (2005) identified two more new diterpenoids forskolin I (1 α, 6 β diacetoxy-7 β, 9 α-dihydroxy-8, 13-epoxylabd-14-en-11-one) and forskolin J (1 α, 9 α-dihydroxy-6 β, 7 β diacetoxy-8, 13-epoxylabd-14-en-11-one) from *C. forskohlii* of Yunnan (Shen and Xu 2005). The labdane diterpene glycosides such as forskoditerpenoside A and B were also obtained from the ethanol extract of the whole plant (Shan et al. 2007). This was the first report on the occurrence of glycosides derived from labdane diterpene in the nature and these compounds showed relaxative effects on isolated guinea pig tracheal spirals. Later, three new minor labdane diterpene glycosides, forskoditerpenoside C, D, and E, and a novel labdane diterpene forskoditerpene A, from the ethanol extract of the whole plant of *C. forskohlii* were isolated (Shan et al. 2008).

A sequence of phyto-products by de Souza and Shah (1988) were identified such as β-configuration of the 7-oxygenated group, i.e., deacetyl forskolin and 9-deoxy forskolin (Bhat et al. 1993), 1,9-dideoxy forskolin (de Souza and Shah 1988), 1,9-dideoxy-7-deacetyl forskolin (Rupp et al. 1985), 7 β-acetoxyl-6 β, 9-α-dihydroxy-8, 13-epoxy-labd-14-en-11-one (Gupta 1988), 6 β, 7 β, 9-α-dihydroxy-8,13 epoxy-labd-14-ene-11-one (Gupta 1988), 6 β,7 β,9α-trihydroxy-8,13-epoxy-labd-14-en-11-one (Akhila et al. 1990) 6 β,dihydroxy-8, 13-epoxy labd-14-en-11-one (Vishwakarma et al. 1988), and 8, 13-epoxy-epoxy-labd-14-en-11-one (Mersinger et al. 1988), respectively. A configuration of the 7-oxygenated group was coleonol B and C, deoxy coleonol, coleonol E (Sen et al. 1992), coleonol D (Tripathi et al. 1995), and coleonol F (Krombholz et al. 1992). The compounds in which, C-11 is not oxygenated have also been reported such as Coleonome (Akhila et al. 1990) and Coleonol (Delpech et al. 1996). Other secondary compounds found in *C. forskohlii* are monoterpenes, monoterpene glycosides, sesquiterpenes, and phenolic glycosides (Ahmed and Merotra 1991; Ahmed and Vishwakarma 1988).

C. forskohlii was reported to contain diterpenoids viz., deactylforskolin, 9-deoxyforskolin, 1,9-deoxyforskolin, 1,9-dideoxy-7- deacetyl forskolin in addition to forskolin (7 β-acetoxy- 8,13-epoxy-1α, 6 β, 9 α-trihydroxylabd-14-en-11-one) in tuber roots extract (Ammon and Kemper 1982; de Souza and Shah 1988; Bhat et al. 1977a, b; Saleem et al. 2006). Forskolin was discovered in 1974 and termed as coleonol (Tandon et al. 1977; Dubey et al. 1981; Soni and Singhai 2012). However, after the identification of other coleonols and diterpenoids, the name of coleonol was later changed to forskolin (Ammon and Kemper 1982).

Misra et al. (1994) reported the presence of 3-decanone, bornyl acetate, β-sesquiphellandrene, and γ-eudesmol as major constituents in essential oil from the roots of 10 genotypes of *C. forskohlii.* Chowdhary and Sharma (1998) found 18 important compounds of which 22% were hydrocarbons and 69% were oxygenated compounds with α-fenchyl acetate and α-pinene as major components. Essential oil extracted from the stem showed chemical constituents like α-pinene, β-caryophyllene, sabinene, β-phellandrene, limonene, α-humulene, α-copaene, and caryophyllene oxide (Kerntopf et al. 2002). Essential oil obtained from *Coleus* plants has attracted the food industry and perfumery due to its attractive fragrance and spicy nature.

Xu et al. (2005) reported six compounds from the roots and identified them as 14-deoxycoleon U, demethylcryptojaponol, α-amyrin, betulic acid, α-cedrol, and β-sitosterol. The compounds namely, α-amyrin and betulic acid were first time isolated from *C. forskohlii.* The hexane extract of *C. forskohlii* roots was analyzed using gas chromatography coupled with mass spectrometry (GC-MS-MS) and identified six major chemical components such as α-cedrene, β-cadinene, citronellal, and two labdane derivatives besides ß-citronellol. These components have been characterized and identified as a rich source for medical and other biological properties from *C. forskohlii* (Murugesan et al. 2012).

The forskolin derivatives viz., Δ5–6-deoxy-7-deacetyl-7-methyl amino carbon forskolin (HIL 568), a potential antiglaucoma agent and 6-(3-dimethylamino propionyl) forskolin hydrochloride (NKH 477), a potential cardiotonic agent were developed (Hosono et al. 1990). Newer compounds are being identified from the root extracts of *C. forskohlii.* Xu et al. (2005) obtained six compounds from the roots of *C. forskohlii* and identified structures as α-amyrin, betulic acid, α-cedrol, dimethyl-cryptojaponol, 14-deoxycoleon U, and β-sitosterol and the compounds viz., α-amyrin and betulic acid were isolated from *C. forskohlii* for the first time.

Shaker et al. (2022) investigated the antimicrobial activities of *C. forskohlii* fractions and identified major active compounds. The antimicrobial assays revealed that ethyl acetate was the most potent fraction and the major abundant metabolite of

C. forskohlii. Thymoquinol-2-O-β-glucopyranoside, syringic acid, methyl 3,4,5-trihydroxybenzoate, and luteolin were isolated herein for the first time.

Wang et al. (2009) isolated 12 compounds from *C. forskohlii* of which chamaecydin, 6-alpha-hydroxydemethylcryptojaponol, and alpha-cedrene were isolated for the first time from *C. forskohlii.* Zhang and Kong (2009) isolated and identified six compounds for the first time from the genus *Coleus*. These include cytochalasin B, N-benzoyl-L-phenylalaninol, 3,6-dibenzyl-2,5-dioxopiperazine, 2-furoic acid, vanillic acid, loliolide and forskolin D. Huang et al. (2011) obtained seven compounds from ethyl acetate fraction of *C. forskohlii* and were identified as lupeol, oleanolic acid, Uvalo, β-sitosterol, colonic acid, demethylcryptojaponol, and coleolic acid. Lupeol and Uvalo were isolated from the genus for the first time.

Chromatographic separation of the *n*-hexane extract of the aerial part of *C. forskohlii* (Syn.: *Plectranthus barbatus*) led to the isolation of five abietane-type diterpenes: dehydroabietane, 5,6-didehydro-7-hydroxy-taxodone, taxodione, 20-deoxocarnosol, and 6α,11, 12,-trihydroxy-7β,20-epoxy-8,11,13-abietatriene (Mothana et al. 2014).

The chemical investigation of ethyl acetate soluble fraction of the ethanol extract of *C. forskohlii* (Syn.: *Plectranthus barbatus*) resulted in six compounds, barbaterpene (2'R-hydroxydocosanoylursa-12-en-3β-ol) and barbatusterol new to the literature along with four known compounds, stigmasterol, sugiol, 11,14-dihydroxy-8,11,13-abietatrien-7-one, and caffeic acid (Amina et al. 2018).

MeOH extract from the leaves of *C. forskohlii* showed in vitro anti-trypanosomal activity (dos Santos et al. 2012). The bioassay-guided fractionation resulted in the isolation of a gallic acid derivative, identified as 1,2,3,4,6-penta-O-galloyl-β-D-glucose (PGG), after a thorough NMR and MS spectral analysis. Finally, this compound was tested against trypomastigote forms of *T. cruzi* and displayed sixfold more effectiveness than the standard drug benznidazole. Musayeib et al. (2020) reported plectrabarbene, a new abietane diterpene from *C. forskohlii* (Syn.: *Plectranthus barbatus*) aerial parts.

4.4 Prospective of Bioengineering of Forskolin

The biosynthesis and production of forskolin using microbial tools are yet to be investigated to overcome the technical artifacts. An effective mechanism needs to be developed based on the existing studies in high quantities. Since Forskolin is an effective pharmacological metabolite, there is a need to improve its production and yield capacity in plant or microbial tools. Establishing the recombinant microbial system may lead to achieve the demand for the industry and may become cost-effective as well as safe drugs.

COLEOUS PHYTOCHEMISTRY

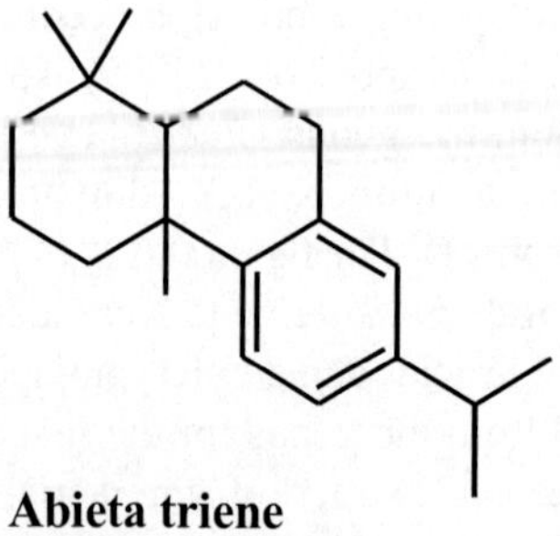

Abieta triene (dehydroabietane)

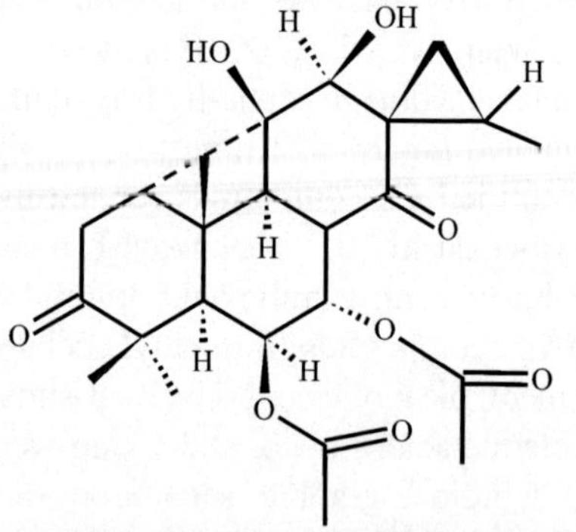

7χ-Acetyl-12-deacetoxy cyclobutatusin

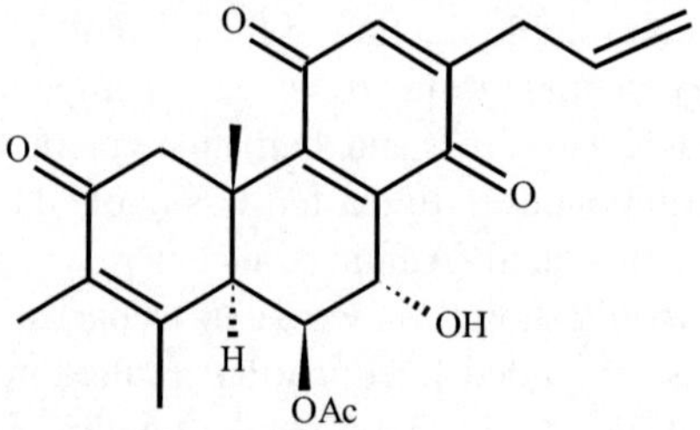

(+)-Allylroyleanone (Plectranthone J)

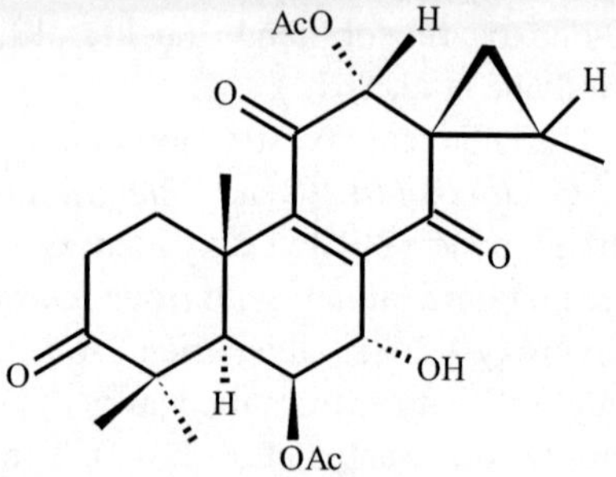

Barbatusin

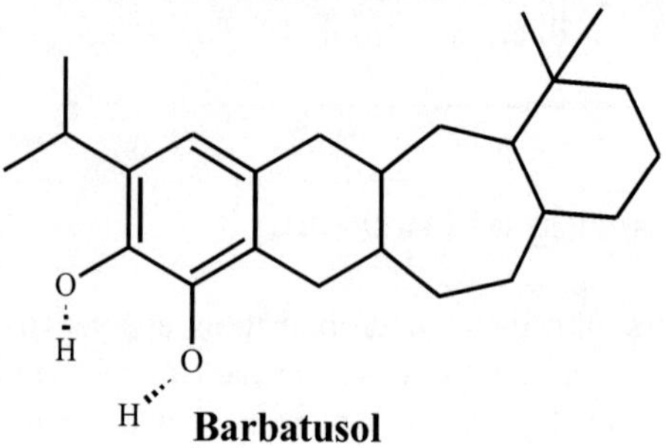

Barbatusol

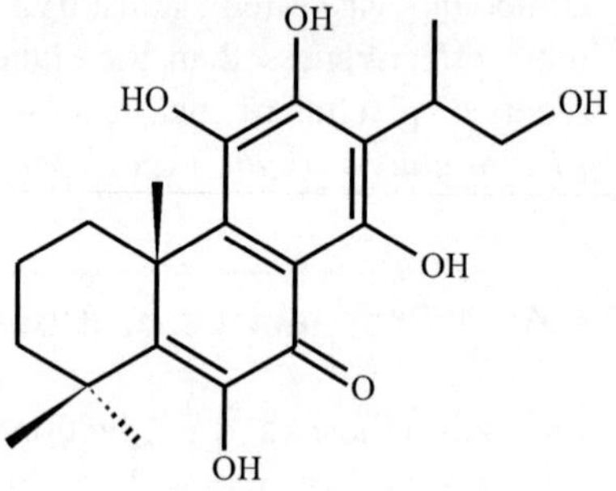

Coleon C

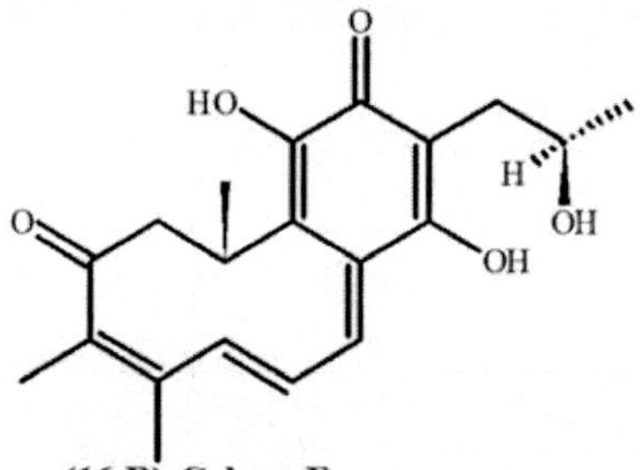

(16 R)-Coleon E

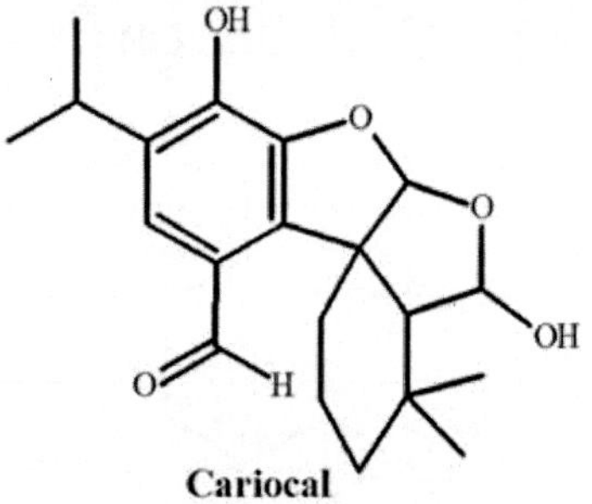

Cariocal

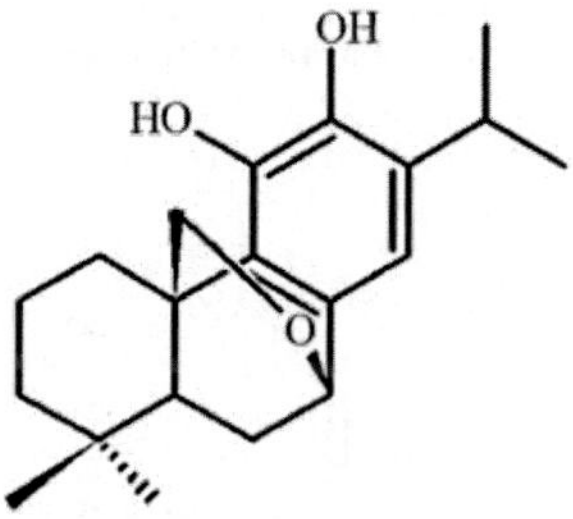

20-Deoxocarnosol

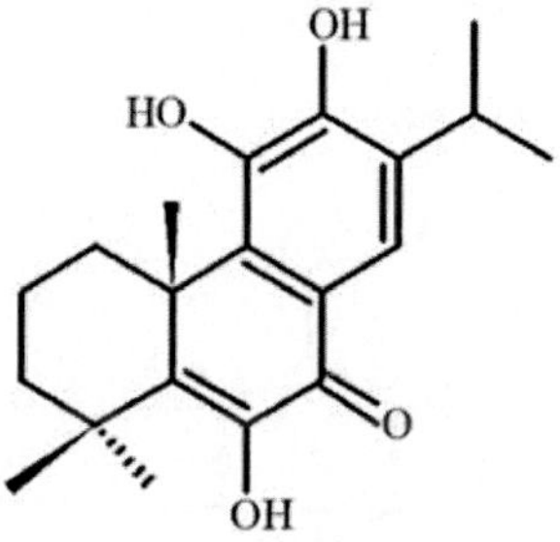

14-Deoxycoleon

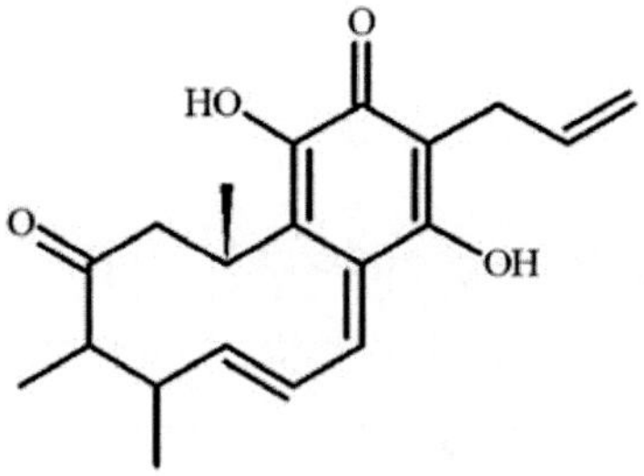

Coleon F

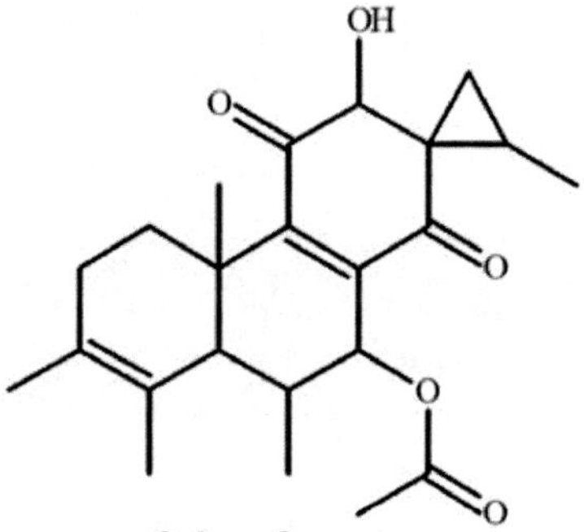

Coleon O

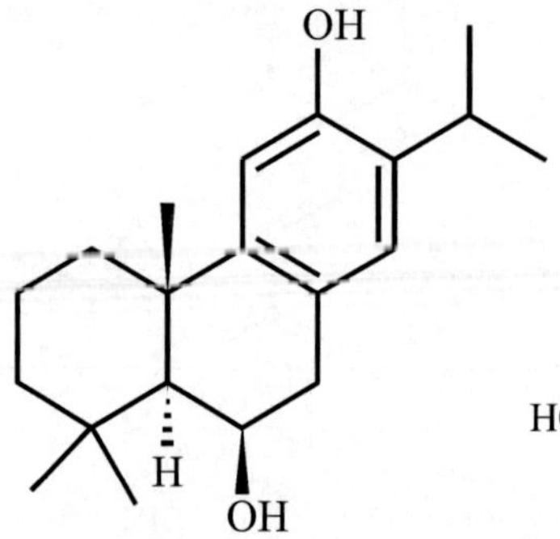

Ferruginol

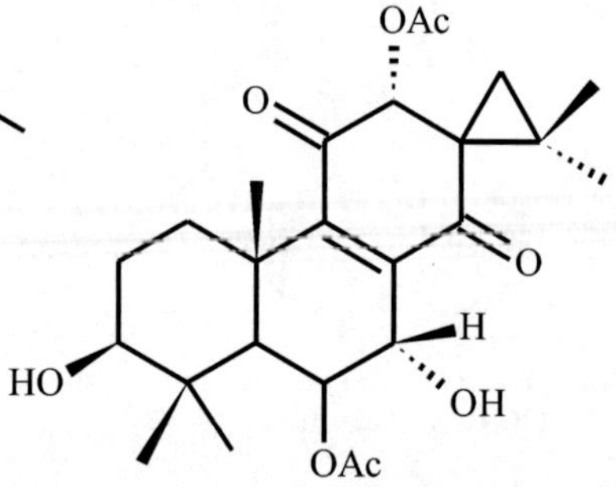

3β-Hydroxy-3-deoxybarbatusin

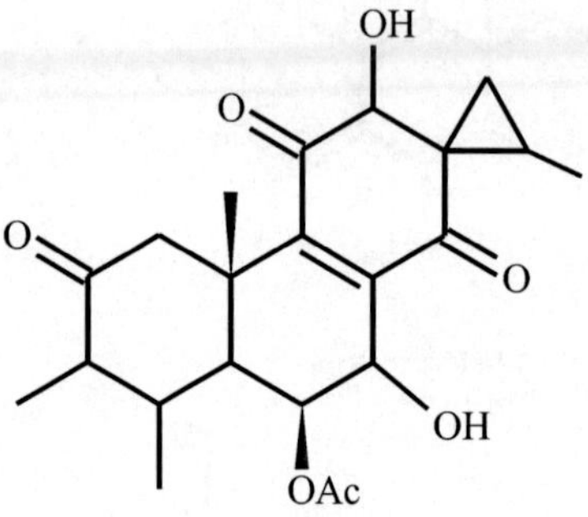

Plectrin

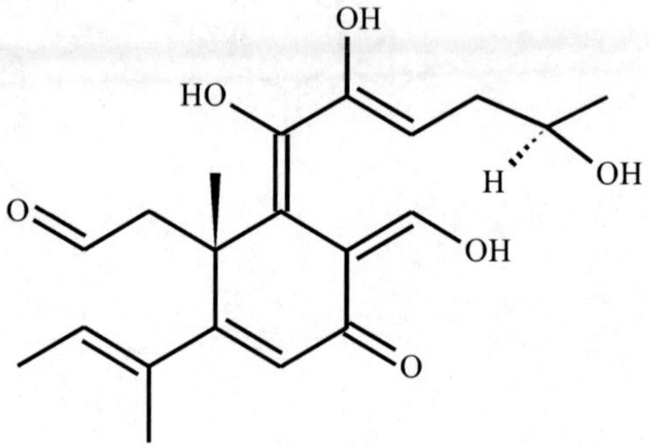

(16R)-Plectrion A

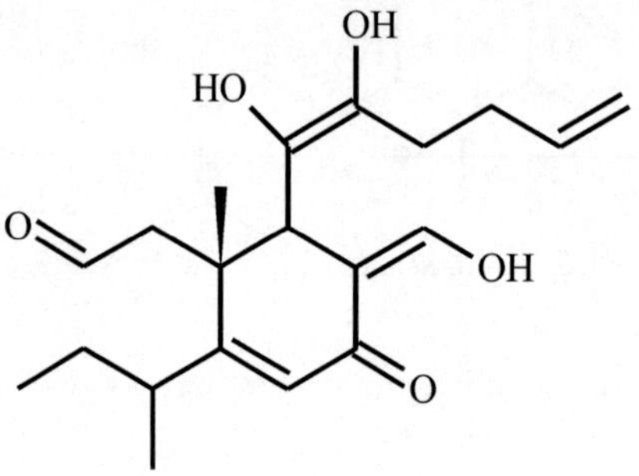

Plecrinon B

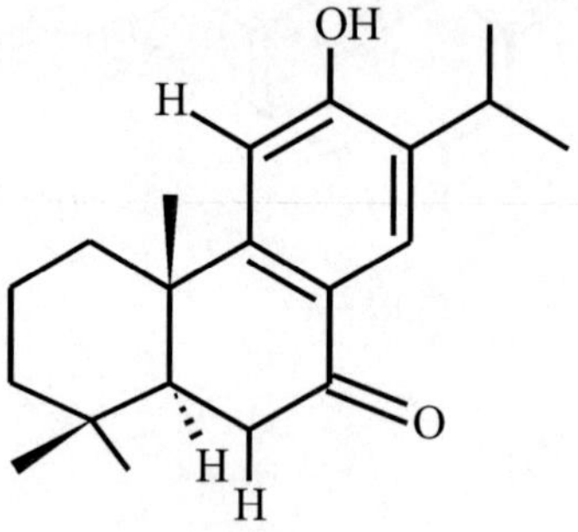

Sugiol

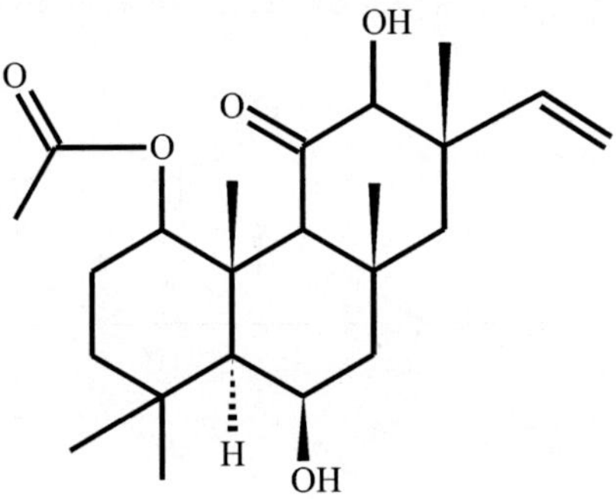

1-Acetoxy Coleosol

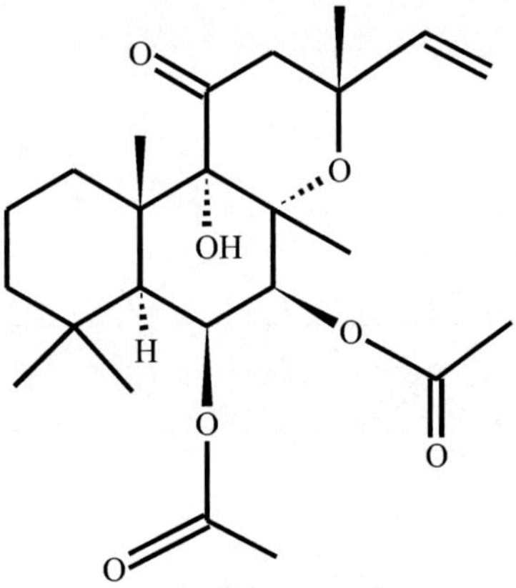

6-Acetoxy-1-deoxy forskolin

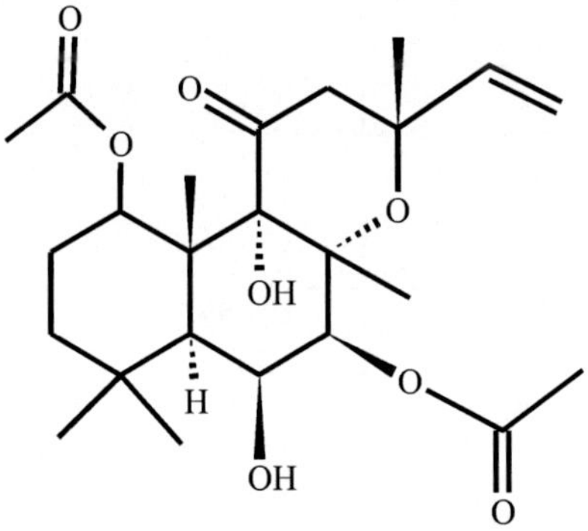

1-Acetylforskolin

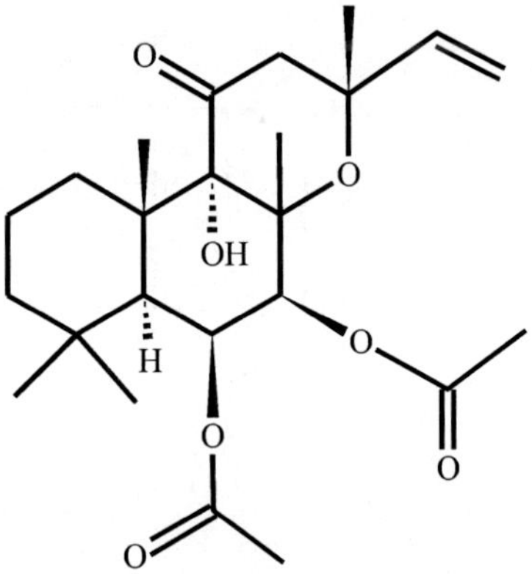

6-Acetyl-1-9-dideoxyforskolin

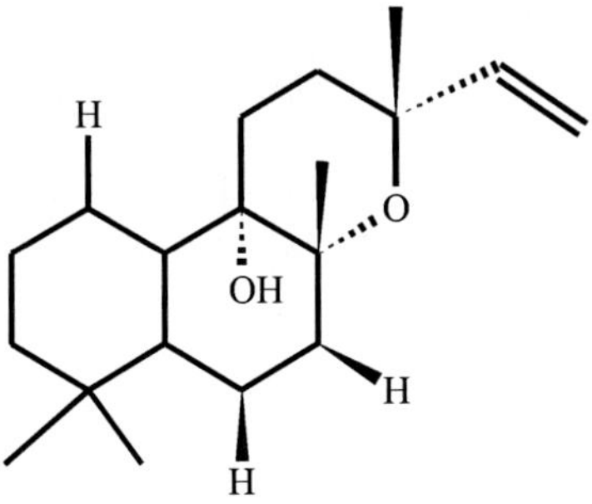

Coleol

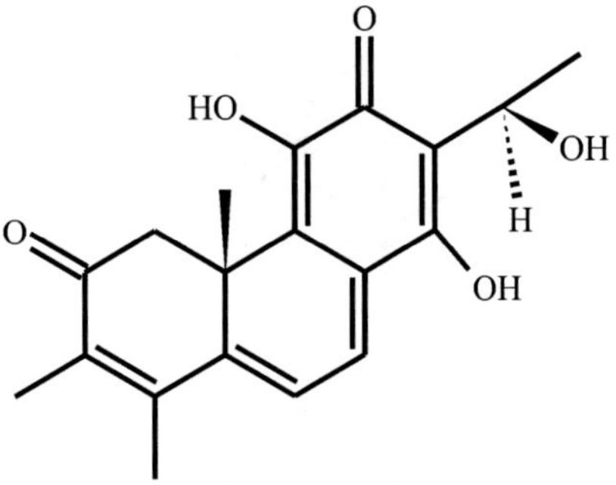

Coleonol E

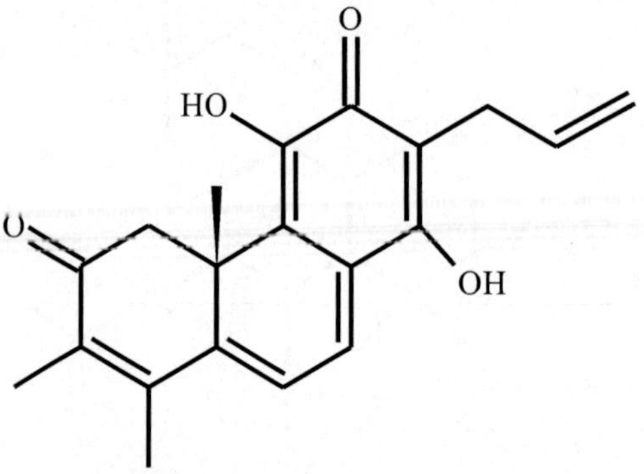

Coleonol F

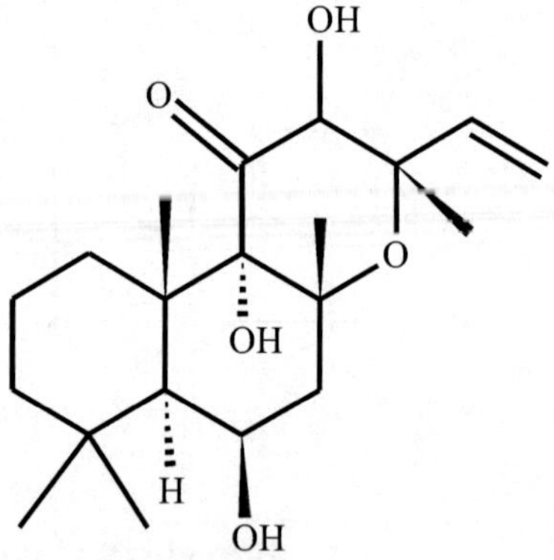

Coleosol

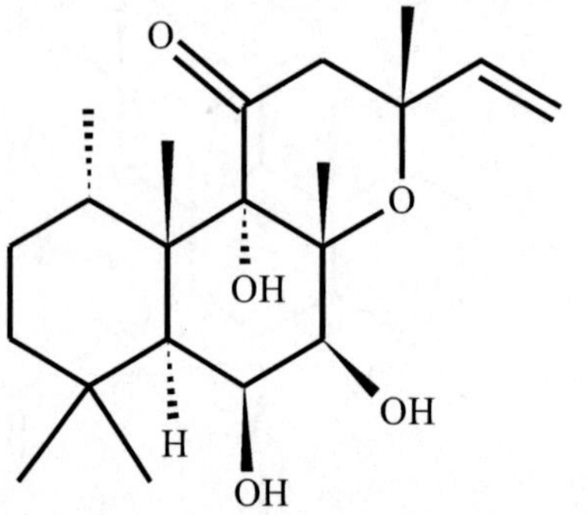

7-Deacetylforskolin

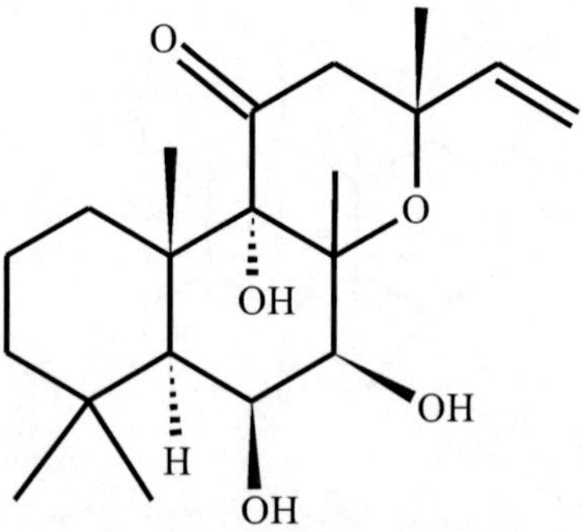

Deacetyl-1-deoxyforskolin

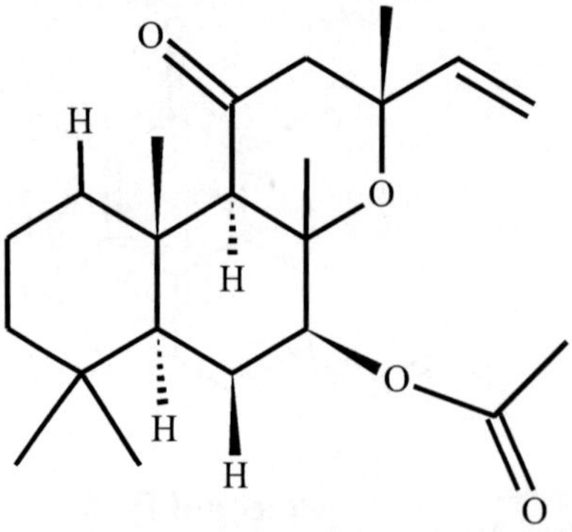

Deoxycoleonol

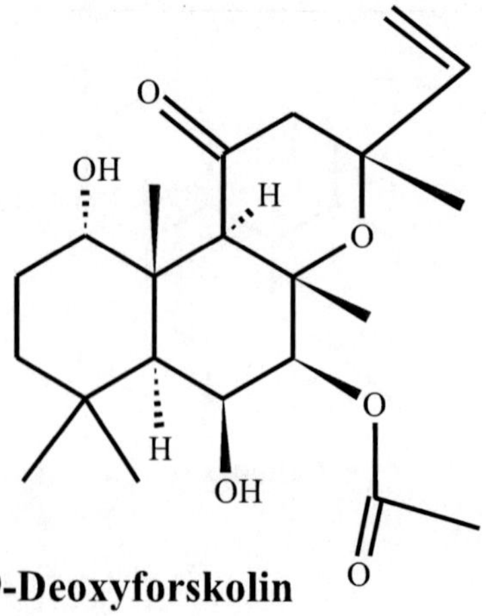

9-Deoxyforskolin

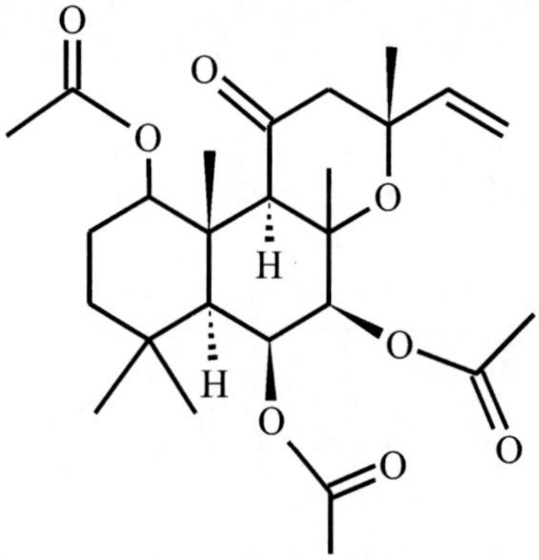

1,6-Diacetoxy-9-deoxyforskolin

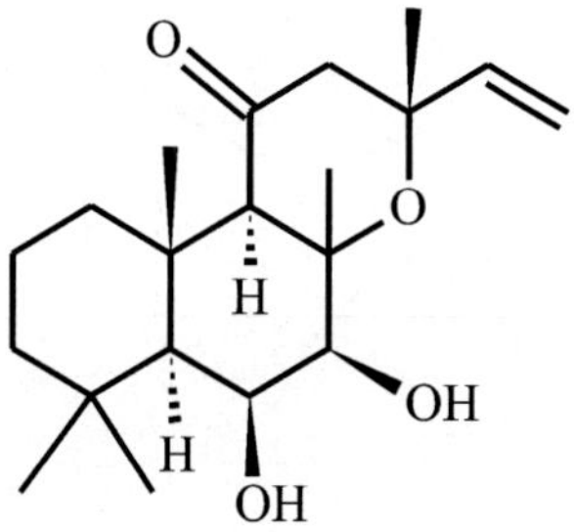

1,9-Dideoxy-7-deacetyl forskolin

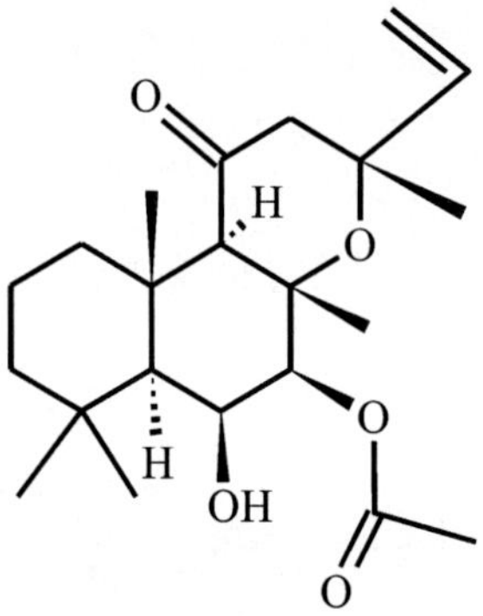

1,9-Dideoxyforskolin

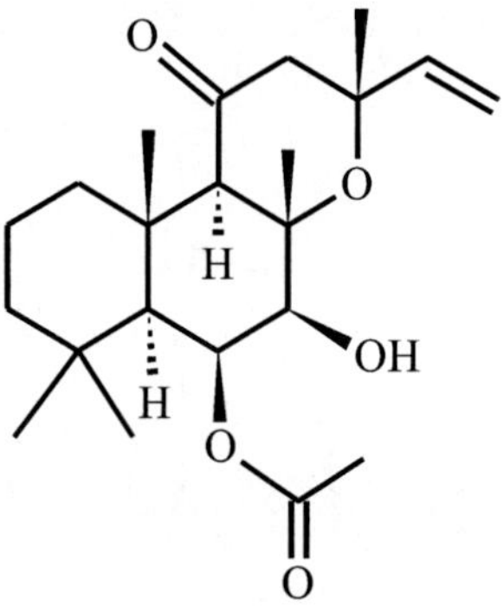

1,9-Dideoxycoleonol B

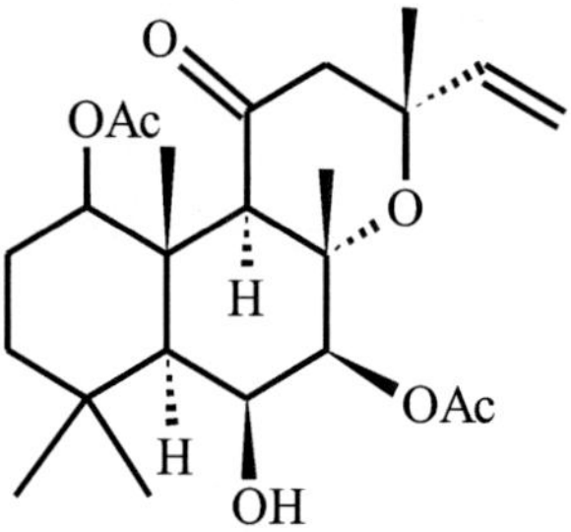

Forskolin E

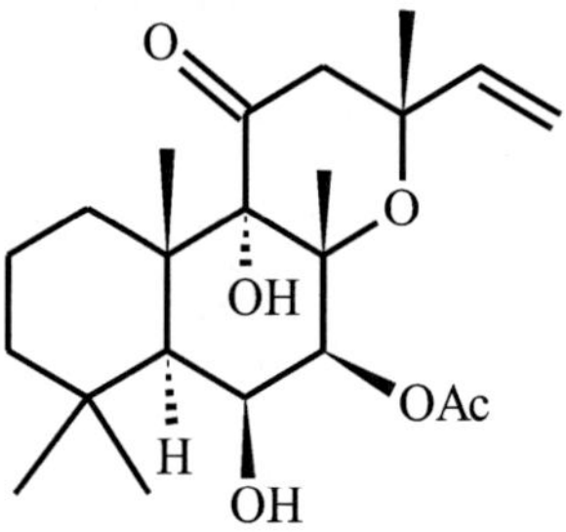

Forskolin F

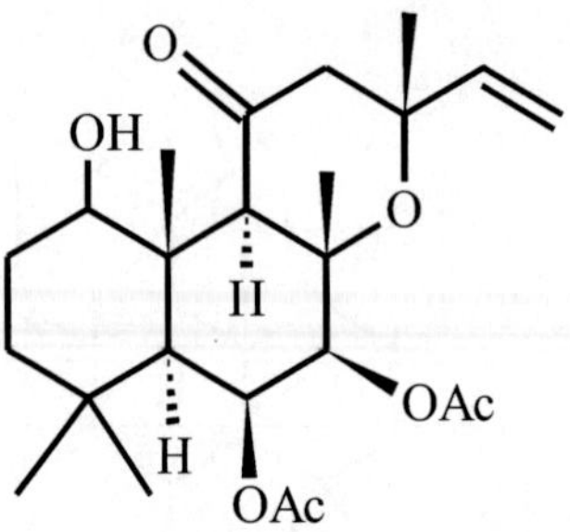

Forskolin G

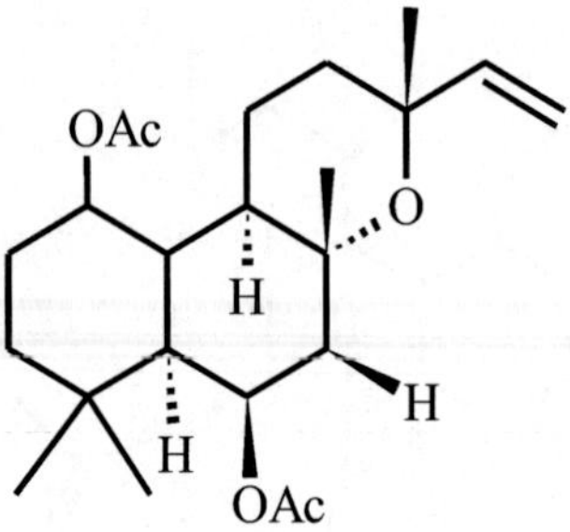

Forskolin H

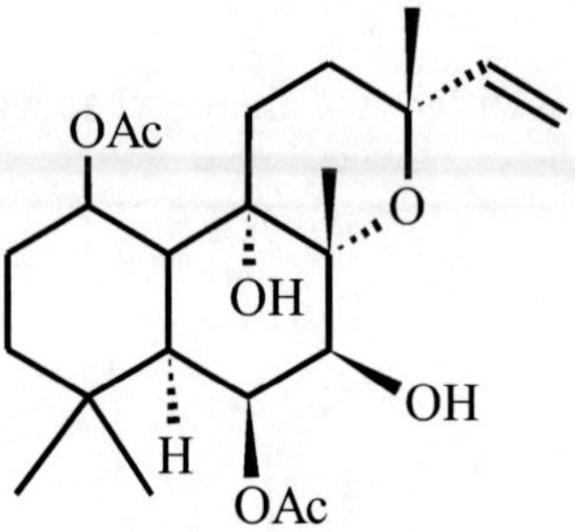

Forskolin I

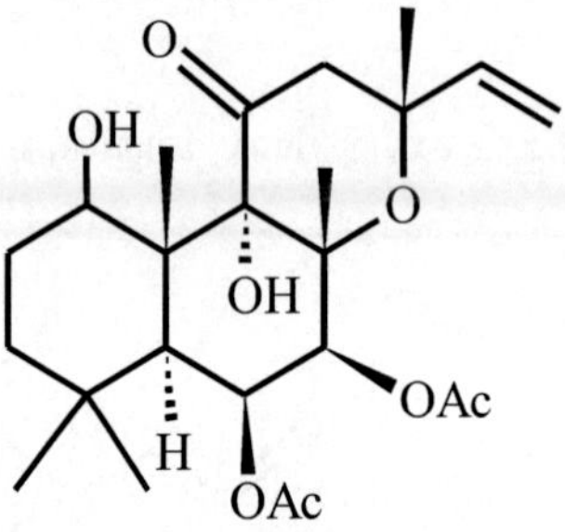

Forskolin J

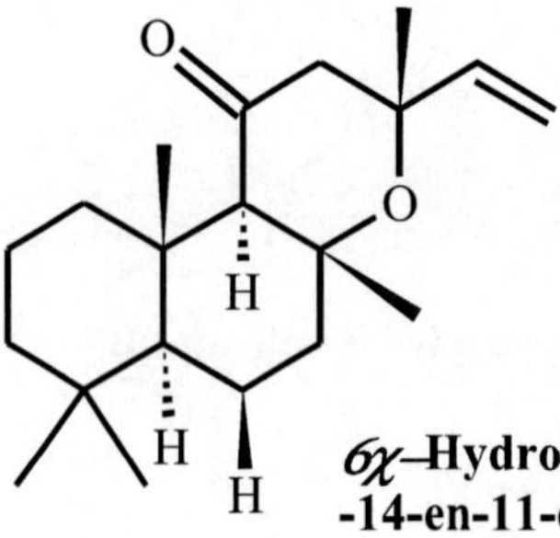

6χ–Hydroxy-8,13-epoxy-1abd -14-en-11-one

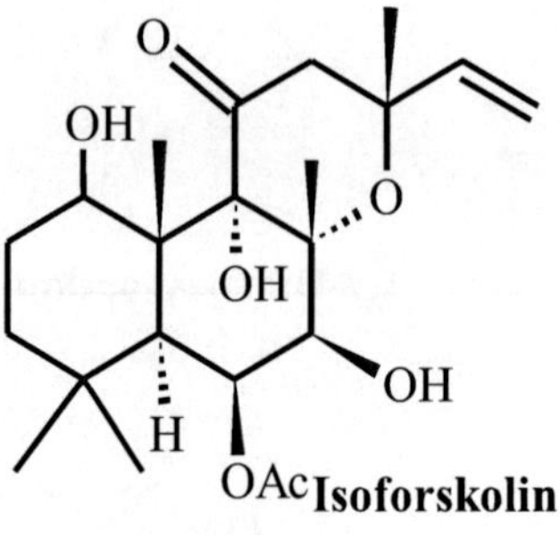

Isoforskolin

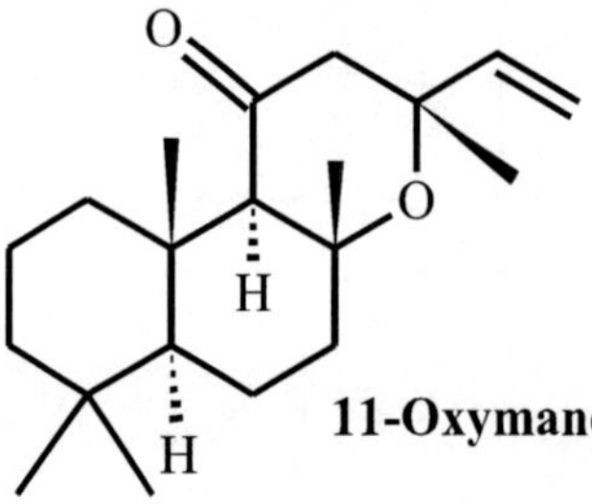

11-Oxymanoyloxide

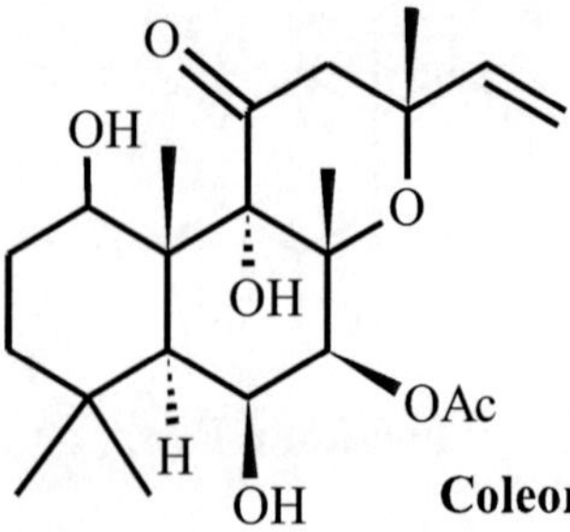

Coleonol

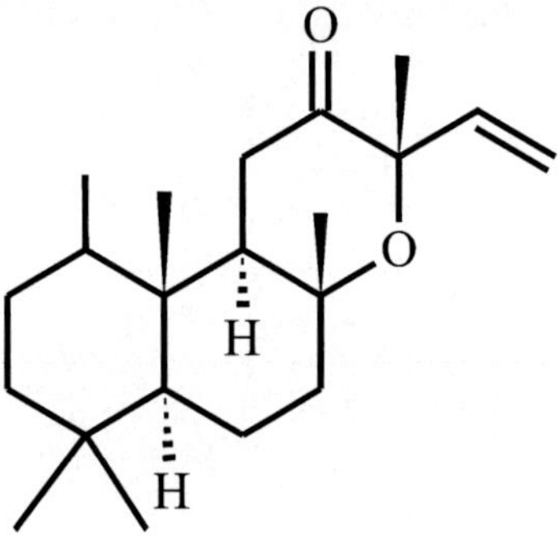

Coleonone

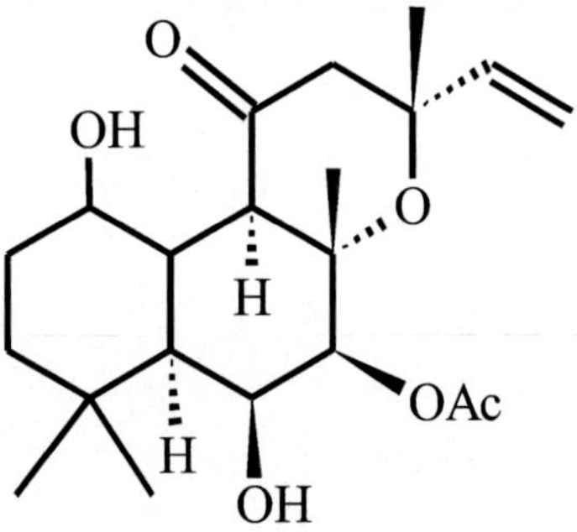

13-Epi-9-deoxycoleonol

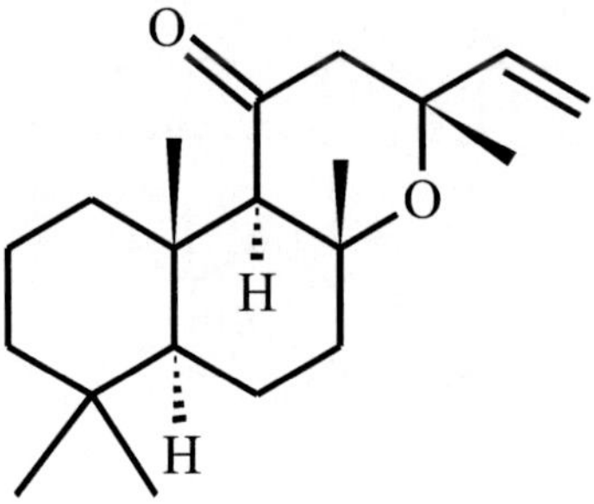

Manoyl oxide

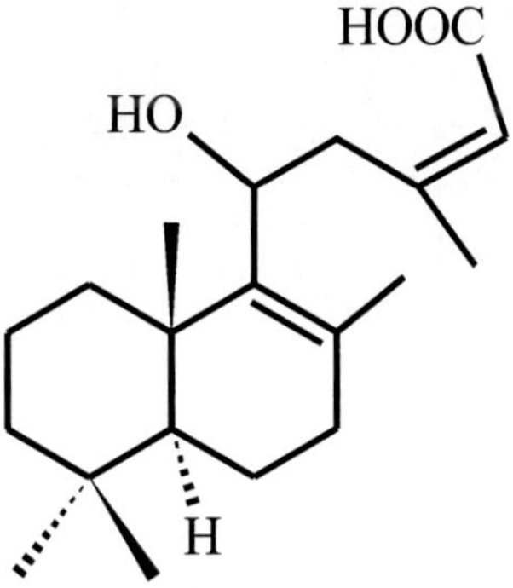

Coleonic acid

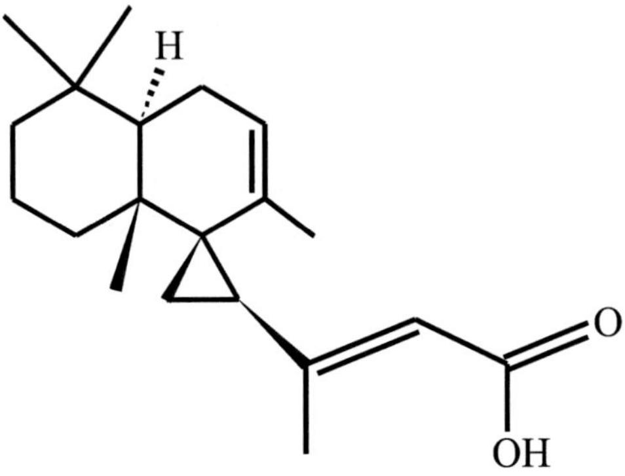

Forskoditerpene A

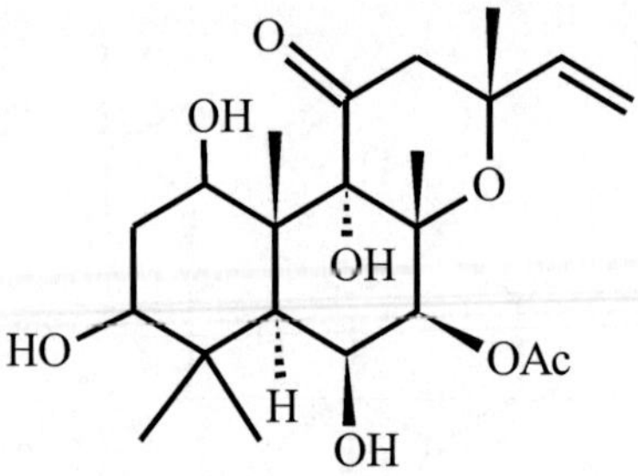

3-Hydroxyforskolin

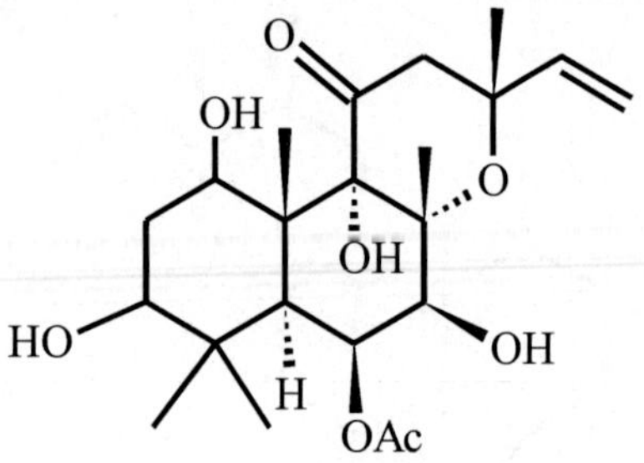

3-Hydroxyisoforskolin

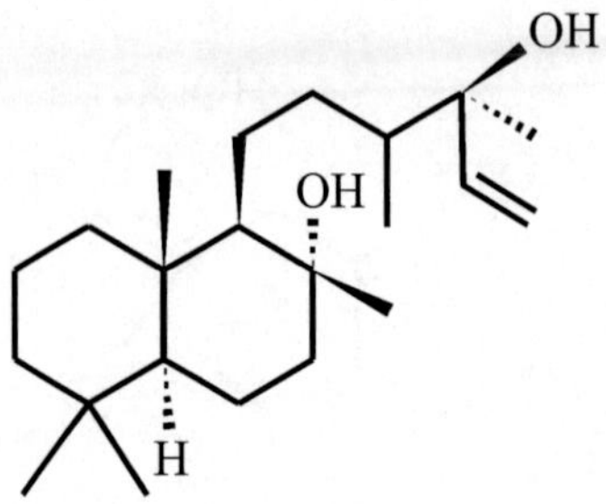

13-Epi-sclareol

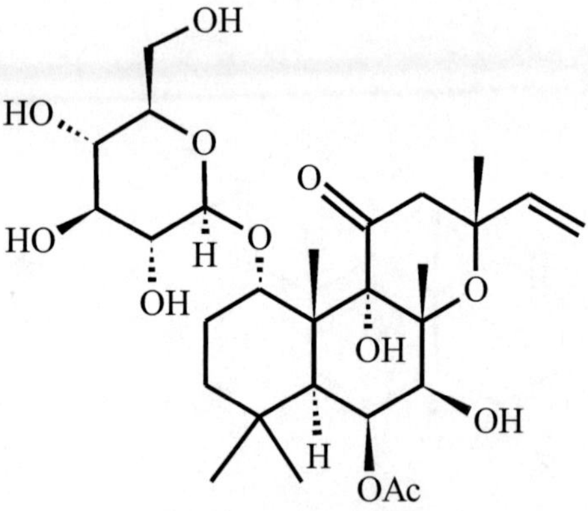

Forskolditerpenoside A

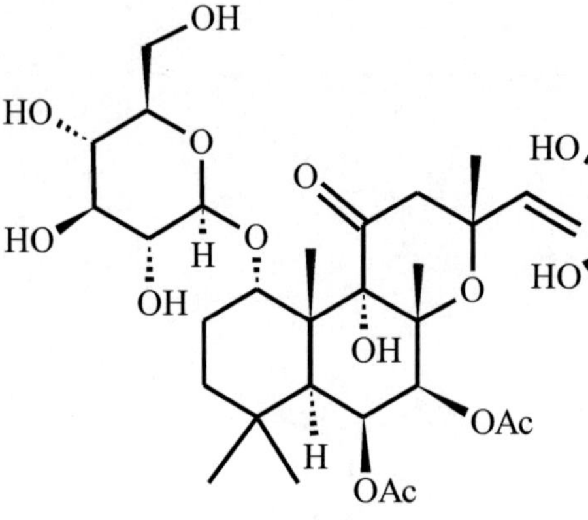

Forskoditerpenoside B

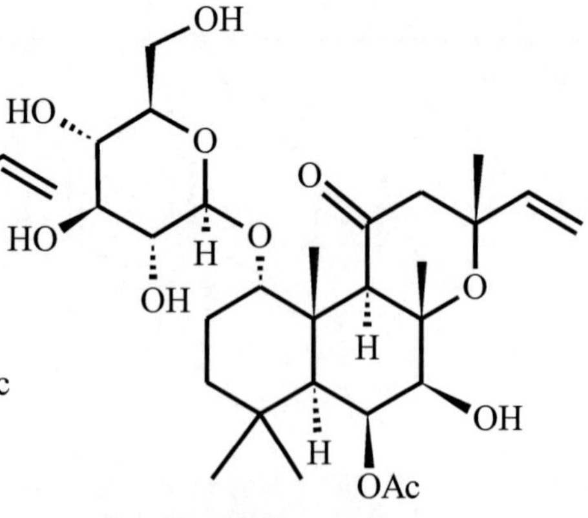

Forskoditerpenoside C

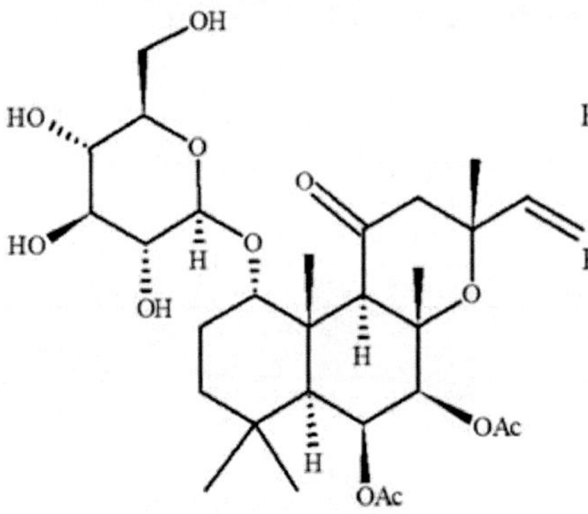

Forskoditerpenoside D

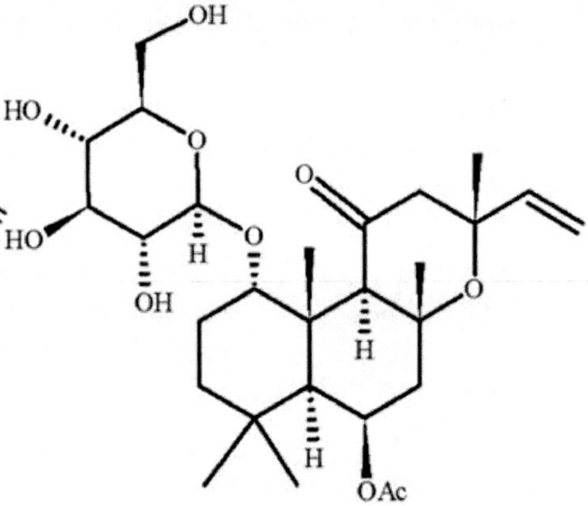

Forskoditerpenoside E

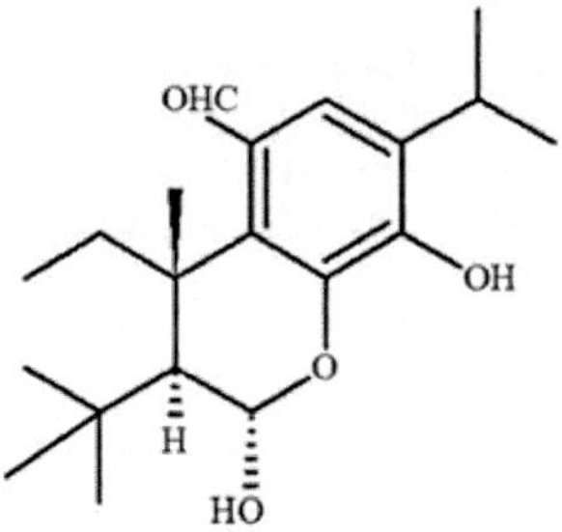

6,7 Seco abietane diterpene I

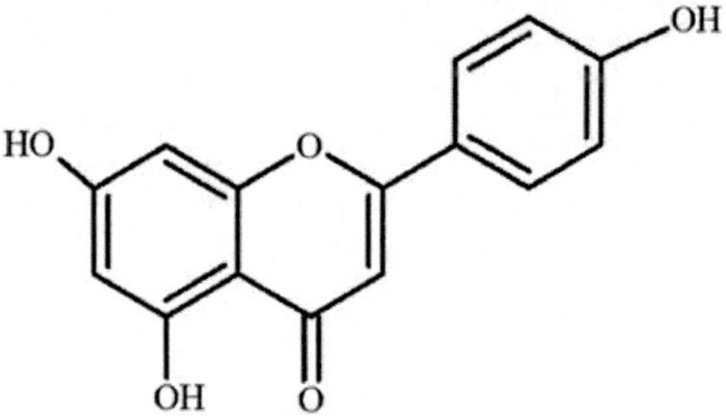

Genkwanin

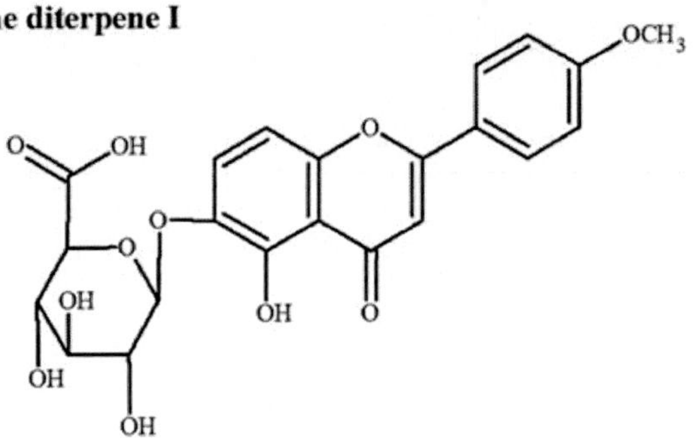

Scutellarein 40-methyl ether 7-O-glucuronide

4.5 Conclusion

The present discussion highlighted the various phytoconstituents of the *C. forskohlii* from leaves, stems, and roots especially diterpenoids with special reference to Forskolin a pharmacologically active ingredient. The suitable methods for obtaining the forskolin from its stems and roots were discussed to understand the convenient procedure for its isolation. The biosynthetic path way gives indication to the propagators to induce high per cent of forskolin during cultivation. This plant has demand over the global markets for its application in different human ailments including cancer. Moreover, forskolin is obtained naturally from this plant without exemption. Hence, the biotechnological approach is essential to obtain the high forskolin lines and necessary to develop high throughput technology for its isolation and processing.

References

Ahmed B, Merotra R (1991) Coleoside-B: a new phenolic glycoside from *Coleus forskohlii*. Pharmazie 46:157–158

Ahmed B, Vishwakarma RA (1988) Coleoside, a monoterpene glycoside from *Coleus forskohlii*. Phytochemistry 27:3309–3310

Akhila A, Rani K, Thakur RS (1990) Biogenetic relationship of polyoxygenated diterpenes in *Coleus forskohlii*. Phytochemistry 29:821–824

Alasbahi RH, Melzig MF (2010) *Plectranthus barbatus*: a review of phytochemistry, ethnobotanical uses and pharmacology. Planta Med 76:753–765

Almeida FC, Lemonica JP (2000) The toxic effects of *Coleus barbatus* B on the different periods of pregnancy in rats. J Ethnopharmacol 73:53–60

Amina M, Al Musayeib NM, Al-Said MS, Al-Zahrani RA, Ibrahim SRM, Mohamed GA (2018) Barbaterpene and barbatusterol, new constituents from *Plectranthus barbatus* growing in Saudi Arabia. Lett Drug Des Discovery 15(8):851–856

Ammon HP, Kemper FH (1982) Ayurveda: 3000 years of Indian traditional medicine. Med Welt 33:148–153

Asada Y, Li W, Terada T, Kuang XZ, Li Q, Yoshikawa T, Hamaguchi S, Namekata I, Tanaka H, Koike K (2012) Labdanetype diterpenoids from hairy root cultures of *C. forskohlii*, possible intermediates in the biosynthesis of forskolin. Phytochemistry 79:141–146

Awasthi P, Gupta AP, Bedi YS, Vishwakarma Ram A, Gandhi SG (2016c) Mannitol stress directs flavonoid metabolism toward synthesis of flavones via differential regulation of two cytochrome P450 monooxygenases in *Coleus forskohlii*. Front Plant Sci 7. https://doi.org/10.3389/fpls.2016.00985

Awasthi P, Lakshmi Jamwal V, Kapoor N, Rasool S (2016a) Homology modeling and docking study of chalcone synthase gene (CfCHS) from *Coleus forskohlii*. Cogent Biol 2:1175332. https://doi.org/10.1080/23312025.2016.1175332

Awasthi P, Mahajan V, Jamwal VL (2016b) Cloning and expression analysis of chalcone synthase gene from *Coleus forskohlii*. J Genet. http://www.ias.ac.in/public/Resources/General/jgen/jgen-15-498-ue.pdf

Awasthi P, Mahajan V, Rather IA, Gupta AP, Rasool S, Bedi YS et al (2015) Plant Omics: isolation, identification, and expression analysis of cytochrome P450 gene sequences from *Coleus forskohlii*. OMICS 19:782–792. https://doi.org/10.1089/omi.2015.0148

Bhat SV, Bajqwa BS, Dornauer H, Do Scusa NJ, Fehlhaber HW (1977a) Structures and stereochemistry of new labdane diterpiniods from *Coleus forskohlii* Briq. Tetrahedron Lett 18:1669–1672

Bhat SV, Bajwa BS, Dornauer H, de Souza NJ, Fehlhaber HW (1977b) Structures and stereochemistry of new labdane diterpenoids from *Coleus forskohlii* Briq. Tetrahedron Lett 19:1669–1672

Bhat SV, Dohadwalla AN, Bajwa BS, Dadkar NK, Dornauer H, de Souza NJ (1983) The antihypertensive and positive inotropic diterpene forskolin: effects of structural modifications on its activity. J Med Chem 26:486–492

Bhat SV, Minale L, Ricco R, Zollo F (1993) Forskolin and congeners. In: Herz W, Kirby GW, Moore RE, Steglich, Tamm CH (eds) Progress in the chemistry of organic natural products. Springer-Verlag, New York, pp 1–60

Bhowal M, Mehta DM (2017) *Coleus forskohlii*: phytochemical and pharmacological profile. Int J Pharm Sci Res 1:3599–3618

Carpy A, Leger JM, Tandon JS, Saxena AK (1991) Molecular and crystal structure of coleonol-B, $C_{22}H_{34}O_7$. Z Kristallogr 194:229–233

Chowdhary AR, Sharma ML (1998) GC-MS investigations on the essential oil from *Coleus forskohlii* Briq. Indian Perfumer 42:15–16

de Albuquerque RL, Kentopff MR, Machado MIL, Silva MGV, de Matos FJA, Morais SM, Braz-Filho R (2007) Abietane diterpenoids isolated from *Plectranthus barbatus* Andrews. Quim Nova 30:1882–1886

De Souza N, Shah V (1988) Forskolin—an adenylate cyclase activating drug from an Indian herb. Econ Med Plant Res 2:1–16

Delpech B, Calvo D, Lett R (1996) Total synthesis of forskolin. Part I. Tetrahedron Lett 37:1015–1018

Demetzos C, Kolocouris A, Anastasaki T (2002) A simple rapid method for the differentiation of C-13 manoyl oxide epimers in biologically important samples using GC-MS analysis supported with NMR spectroscopy and computational chemistry results. Bioorg Med Chem Lett 12:3605–3609

dos Santos RT, Hiramoto LL, Lago JH, Sartorelli P, Tempone AG, Pinto EG (2012) Anti-trypanosomal activity of 1,2,3,4,6-penta-O-galloyl-β-D-glucose isolated from *Plectranthus barbatus* Andrews (Lamiaceae). Quim Nova 35:11. https://doi.org/10.1590/S0100-40422012001100025

Dubey MP, Srimal RC, Nityanand S, Dhawan BN (1981) Pharmacological studies on colenol, a hypotensive diterpene from *Coleus forskohlii*. J Ethnopharmacol 3:1–13

Eisenreich W, Schwarz M, Cartayrade A, Arigoni D, Zenk MH, Bacher A (1998) The deoxyxylulose phosphate pathway of terpenoid biosynthesis in plants and microorganisms. Chem Biol 5:R221–R233

Englund E, Andersenranberg J, Miao R, Hamberger B, Lindberg P (2015) Metabolic engineering of *Synechocystis* sp. PCC 6803 for production of the plant diterpenoid manoyl oxide. ACS Synth Biol 4:1270–1278

Engprasert S, Taura F, Kawamukai M, Shoyama Y (2004) Molecular cloning and functional expression of geranyl-geranyl-pyrophosphate synthase from *Coleus forskohlii* Briq. BMC Plant Biol 4:18. https://doi.org/10.1186/1471-2229-4-18

Engprasert S, Taura F, Shoyama Y (2005) Molecular cloning, expression and characterization of recombinant1-deoxy-D-xylulose-5-phosphate reductoisomerase from *Coleus forskohlii* Briq. Plant Sci 169:287–294. https://doi.org/10.1016/j.plantsci.2004.07.017

Falé PL, Borges C, Madeira PJA, Ascensão L et al (2009) Rosmarinic acid, scutellarein 4′-methyl ether 7-O-glucuronide and (16S)-coleon E are the main compounds responsible for the antiacetylcholinesterase and antioxidant activity in herbal tea of *Plectranthus barbatus* ("falso boldo"). Food Chem 114(3):798–805. https://doi.org/10.1016/j.foodchem.2008.10.015

Feng S, Wang L, Voravuthikunchai S, Liang Z, Liu A, Bao F (2016) Isoforskolin from native plant *Coleus forskohlii* of Yunnan, China plays multiple biological roles. Open J Immunol 6:63–66. https://doi.org/10.4236/oji.2016.62006

Gabetta B, Zini G, Danieli B (1989) Minor diterpenoids of *Coleus forskohlii*. Phytochemistry 28: 859–862

Gao Y, Liu F, Fang L, Cai R, Zong C, Qi Y (2014) Genkwanin inhibits proinflammatory mediators mainly through the regulation of miR-101/MKP-1/MAPK pathway in LPS-activated macrophages. PLoS One 9:e96741. https://doi.org/10.1371/journal.pone.0096741

Gupta R (1988) Genetic resources of medicinal plants. Indian J Plant Genet Resour 1:98–102

Hashimoto S, Sakata S, Sonegawa M, Ikegami S (1988) A total synthesis of (.+−.)-forskolin. J Am Chem Soc 110(11):3670–3672

Hosono M, Takahira T, Fujita A, Fujihara R, Ishizuka O, Ohio I, Tatee T, Nakamura K (1990) Cardiovascular effects of NICH 477, a novel potent water soluble forskolin derivative. Eur J Pharmacol 183:2110–2011

Huang DH, Yang YF, Ali LQ, Lu Y, Wu HZ (2011) Studies on the chemical constituents of *Coleus forskohlii* transplanted in Tongcheng and their antitumor activity. Zhong Yao Cai 34(3): 375–378

Inamdar PK, Dornauer H, de Souza NJ (1980) GLC method for assay of forskolin, a novel positive inotropic and blood pressure-lowering agent. J Pharm Sci 69(12):1449–1451

Inamdar PK, Kanitakar PV, Reden J, de Souza NJ (1984) Quantitative determination of forskolin by TLC and HPLC. Panta Med 50(1):30–34

Jauhari PK, Katti SB, Tandon JS, Dhar MM (1978) Coleosol – a new diterpene from *Coleus forskohlii*. Indian J Chem Sect B Org Chem Incl Med Chem 16B:1055–1057

Jeon YM, Kim BG, Ahn JH (2009) Biological synthesis of 7-O-methyl apigenin from naringenin using *Escherichia coli* expressing two genes. J Microbiol Biotechnol 19:491–494. https://doi.org/10.4014/jmb.0807.402

Jin QD, He BH (1998) Minor constituents from *Coleus forskohlii*. Acta Bot Yunn 20:469–473

Jin QD, Xie XH, Mu QZ (1990) Study on the chemical constituents from *Coleus forskohlii* Briq. Nat Prod Res Dev 2:6–9

Ju H, Zhang C, Lu W (2021) Progress in heterologous biosynthesis of forskolin. J Ind Microbiol Biotechnol 48:kuab009

Katti SB, Jauhari PK, Tandon JS (1979) New diterpenes from *Coleus forskohlii*: structures of the diterpenes, coleonol-D, coleol and coleonone. Indian J Chem Sect B Org Chem Incl Med Chem 17B:321–323

Kavitha C, Rajamani K, Vadivel E (2010) *Coleus forskohlii*: a comprehensive review on morphology, phytochemistry and pharmacological aspects. J Med Plant Res 4:278–285

Kelecom A (1983a) An abietane diterpene from the labiate *Coleus barbatus*. Phytochemistry 23(8): 1677–1678

Kelecom A (1983b) Isolation, structure determination, and absolute configuration of barbatusol, a new bioactive diterpene with a rearranged abietane skeleton from the labiate *Coleus barbatus*. Tetrahedron 39:3603–3608

Kelecom A, dos Santos TC (1985) Cariocal, a new seco-abietane diterpene from the labiate *Coleus barbatus*. Tetrahedron Lett 26:3659–3662

Kelecom A, Dos Santos TC, Medeiros WLB (1986) On the structure and absolute configuration of (−)-20-deoxocarnosol. An Acad Bras Cienc 58:53–59

Kelecom A, Dos Santos TC, Medeiros WLB (1987) Secoabietane diterpenes from *Coleus barbatus*. Phytochemistry 26:2337–2340

Kerntopf MR, de Albuquerque RL, Machado MIL, Matos FJA, Craveiro AA (2002) Essential oil from leaves, stems and roots of *Plectranthus barbatus* Andr. (Labiatae) grown in Brazil. Brazilian J Essential Oil Res 14:101–102

Khandelwal Y, Jotwani PK, Inamdar PK, de Souza NJ, Rupp RH (1989) Isolation, structure elucidation, and synthesis of 1-deoxyforskolin. Tetrahedron 45:763–766

Krombholz R, Mersinger R, Kreis W, Reinhard E (1992) Production of forskolin by axenic *Coleus forskohlii* roots cultivated in shake flasks and 20-l glass jar bioreactors. Planta Med 58:328–333

Kubo I, Matsumoto T, Tori M, Asakawa Y (1984) Structure of plectrin, an aphid antifeedant diterpene from *Plectranthus barbatus*. Chem Lett 9:1513–1516

Li S, Yang QR, Wang XM, Zou GA, Liu YW (2006) Chemical constituents of *Coleus forskohlii* replanted to Tongcheng. Zhongcaoyao 37:824–826

Liu C, Yao W, Meng K, Liu Y, Zhang S, Zeng S, Lu S, Li H (1992) Isolation and purification of forskolin and identification of its chemical structure and biological activity. Shengwu Huaxue Yu Shengwu Wuli Jinzhan 19:74–76

Liu Y, Wang XM, Wu H (2007) Main components of *Coleus forskohlii* extract and relevant extraction method. Chinese Patent CN 1944384 A

Mathela DK, Kharkwal HB, Mathela CS (1986) Terpenoids of roots of *Coleus forskohlii*. Fitoterapia 57:299–301

Mersinger R, Dornauer H, Reinhard E (1988) Formation of forskolin by suspension culture of *Coleus forskohlii*. Planta Med 54:200–204

Misra LN, Tyagi BR, Ahmad A, Bahl JR (1994) Variablity in the chemical composition of the essential oil of *Coleus forskohlii* genotypes. J Essent Oil Res 6:243–247

Mothana RA, Al-Said MS, Al-Musayeib NM, El Gamal AA, Al-Massarani SM, Al-Rehaily AJ, Abdulkader M, Maes L (2014) In vitro antiprotozoal activity of abietane diterpenoids isolated from *Plectranthus barbatus* Andr. Int J Mol Sci 15(5):8360–8371

Murugesan S, Rajeshkannan C, Sumathi R, Manivachakam P, Suresh Babu D (2012) Bioactivity of root hexane extract of *Coleus forskohlii* Briq. Labiatae GC/MS characterization and identification. Eur J Exp Biol 2(5):1469–1473

Musayeib NMA, Amina M, Al-Hamoud GA, Mohamed GA, Ibrahim SRM, Shabana S (2020) Plectrabarbene, a new abietane diterpene from *Plectranthus barbatus* aerial parts. Molecules 25: E2365

Painuly P, Katti SB, Tandon JS (1979) Diterpenes from *Coleus forskohlii*: structures of coleonol-E and coleonol-F. Indian J Chem Sect B Org Chem Incl Med Chem 18B:214–216

Pateraki I, Andersen-Ranberg J, Hamberger B, Heskes AM, Martens HJ, Zerbe P, Bach SS, Moller BL, Bohlmann J, Hamberger B (2014) Manoyl oxide (13R), the biosynthetic precursor of forskolin, is synthesized in specialized root cork cells in *C. forskohlii*. Plant Physiol 164: 1222–1236

Pateraki I, Andersen-Ranberg J, Jensen NB, Wubshet SG, Heskes AM, Forman V, Hallstroom B, Hamberger B, Motawia MS, Olsen CE, Staerk D, Hansen J, Moller BL, Hamberger B (2017) Total biosynthesis of the cyclic AMP booster forskolin from *C. forskohlii*. elife 6:e23001

Prakash O, Roy R, Tandon JS, Dhar MM (1988) Carbon-13 and proton two-dimensional NMR study of diterpenoids of *Coleus forskohlii*. Magn Reson Chem 26:117–119

Rajangam J (2005) Studies on the effect of planting methods and growth regulators on growth, tuber development, yield and quality, and standardization of post-harvest technology of coleus (*Coleus forskohlii* Briq.). Ph.D. Thesis, Tamil Nadu Agricultural University, Coimbatore, T.N. India

Rana PS, Saklani P, Chandel C (2021) Analyzing the effects of altitudes on the metabolic diversity and forskolin content in *Coleus forskohlii* roots by HPTLC and HPLC. Plant Biosyst. https://doi.org/10.1080/11263504.2020.1857865

Rohmer M, Knani M, Simonin P, Sutter B, Sahm H (1993) Isoprenoid biosynthesis in bacteria: a novel pathway for the early steps leading to isopentenyl diphosphate. Biochem J 295:517–524

Rohmer M, Seeman M, Horbach S, Bringer-Mayer S, Sahm H (1996) Glyceraldehyde 3-phosphate and pyruvate as precursors of isoprenic units in an alternative non-mevalonate pathway for terpenoid biosynthesis. J Am Chem Soc 118:2564–2566

Roy R, Mishra A, Varma N, Tandon JS, Saux M, Carpy A (1993) Minor diterpenes from *Coleus forskohlii*. Phytochemistry 34:1577–1580

Ruedi P (1986) Novel diterpenoids from leaf glands of *Plectranthus barbatus* (Labiatae). The absolute configuration of the 2-hydroxypropyl group in coleon E. Helv Chim Acta 69:972–984

Ruedi P, Eugster CH (1972) Structure of coleon E, a new diterpenoid methylenequinone from *Coleus barbatus* species (Labiatae). Helv Chim Acta 55:1994–2014

Ruedi P, Eugster CH (1973) Structure of coleon F. Helv Chim Acta 56:1129–1132

Rupp RH, de Souza NJ, Dohadwalla AN (1985) *Forskolin*: its chemical, biological and medical potential. Hoechst India Limited, Bombay

Saleem AM, Dhasan PB, Rafiullah MRM (2005) Isolation of forskolin from stem of *Coleus forskohlii*. Pharmacogn Mag 1(3):89–92

Saleem AM, Dhasan PB, Rafiullah MRM (2006) Simple and rapid method for the isolation of forskolin from *Coleus forskohlii* by charcoal column chromatography. J Chromatogr A 1101(1–2):313–314. https://doi.org/10.1016/j.chroma.2005.11.032

Sashidhara KV, Rosaiah JN, Kumar A, Bid HK, Konwar R, Chattopadhyay N (2007) Cell growth inhibitory action of an unusual labdane diterpene, 13- epi-sclareol in breast and uterine cancers in vitro. Phytother Res 21:1105–1108

Schaneberg BT, Khan IA (2003) Quantitative analysis of forskolin in *Coleus forskohlii* (Lamiaceae) by reversed-phase liquid chromatography. J AOAC Int 86(3):467–470

Schultz C, Bossolani MP, Torres LMB, Lima-Landman MTR, Lapa AJ, Souccar C (2007) Inhibition of the gastric H+, K+-ATPase by plectrinone A, a diterpenoid isolated from *Plectranthus barbatus* Andrews. J Ethnopharmacol 111:1–7

Sen J, Sharma AK, Sahu NP, Mahato SB (1992) Production of forskohlin in *in vitro* cultures of *Coleus forskohlii*. Planta Med 58:324–327

Shah V, Bhat SV, Bajwa BS, Dornauer H, de Souza NJ (1980) The occurrence of forskolin in the Labiatae. Planta Med 39:183–185

Shaker KH, Yonus MA, Ibrahim MA, Kilany M, Wiggers F (2022) Antimicrobial activity of *Coleus forskohlii* fractions, isolation and characterization of phenolic compounds. Curr Bioact Compd 18(1):e160921190769

Shan Y, Wang X, Zhou X, Kong L, Niwa M (2007) Two minor diterpene glycosides and an eudesman sesquiterpene from *Coleus forskohlii*. Chem Pharm Bull 55:376–381

Shan Y, Xu L, Lu Y, Wang X, Zheng Q, Kong L, Niwa M (2008) Diterpenes from *Coleus forskohlii* (Willd.) Broq. (Labiatae). Chem Pharm Bull 56:52–56

Shan YP, Kong LY (2006) Isolation and identification of terpenes from *Coleus forskohlii*. Chin J Nat Med 4:271–274

Shan YP, Wang XB, Kong LY (2006) Forskolin G. Acta Crystallogr Sect E Struct Rep Online E62: o2408–o2410. http://journals.iucr.org/e/issues/2006/06/00/hk2036/index.html

Shen YH, Xu YL (2005) Two new diterpenoids from *Coleus forskohlii*. J Asian Nat Prod Res 7: 811–815

Shen YH, Yao CS, Xu YL (2002) New diterpenoids from *Coleus forskohlii*. Chin Chem Lett 13: 740–743

Singh S, Painuly P, Tandon JS (1984) Diterpenes from *Coleus forskohlii*: stereochemistry of the carbonyl chromophore. Indian J Chem Sect B Org Chem Incl Med Chem 23B:952–955

Soni H, Singhai AK (2012) Recent updates on the genus *Coleus*: a review. Asian J Pharm Clin Res 5:12–17

Sutardja LC, Jensen PE, Moller BL, Nielsen AZ, Gnanasekaran T, Walby SL (2019) Heterologous Production of Forskolin in the Cyanobacterium *Synechocystis* PCC 6803, International Conference on Plant Synthetic Biology and Bioengineering, AIChE Academy. https://www.aiche.org/academy/conferences/international-conference-on-plant-synthetic-biology-and-bioengineering/2019/proceeding

Tandon JS, Dhar MM, Ramakumar S, Venkatesan K (1977) Structure of colenol, a biologically active diterpene from *Coleus forskohlii*. Indian J Chem 15:880–883

Tandon JS, Jauhari PK, Singh RS, Dhar MM (1978) Structures of three new diterpenes, coleonol B, coleonol C and deoxycoleonol isolated from *Coleus forskohlii*. Indian J Chem Sect B Org Chem Incl Med Chem 16B:341–345

Tandon JS, Katti SB, Ruedi P, Eugster CH (1979) Crocetin-dialdehyde from *Coleus forskohlii* Briq. Labiatae Helv Chem Acta 62:2706–2707

Tandon JS, Roy R, Balachandran S, Vishwakarma RA (1992) Epi-deoxycoleonol, a new antihypertensive labdane diterpenoid from *Coleus forskohlii*. Bioorg Med Chem Lett 2:249–254

Tripathi CKM, Basu SK, Jain S, Tandon JS (1995) Production of coleonol (forskolin) by root callus cells of plant *Coleus forskohlii*. Biotechnol Lett 17:423–426

Vishwakarma RA, Tyagi BR, Ahmed B, Hussain A (1988) Variation in forskolin content in the roots of *Coleus forskohlii*. Planta Med 54:471–472

Wang AHJ, Paul IC, Zelnik R, Lavie D, Levy EC (1974) Structure and stereochemistry of cyclobutatusin, a diterpenoid containing a four-membered ring. J Am Chem Soc 96:580–581

Wang AHJ, Paul IC, Zelnik R, Mizuta K, Lavie D (1973) Structure and absolute stereochemistry of the diterpenoid barbatusin. J Am Chem Soc 95:598–600

Wang K, Ohnuma S (1999) Chain-length determination mechanism of isoprenyl diphosphate synthases and implications for molecular evolution. Trends Biochem Sci 24(11):445–451

Wang YQ, Ma JP, Pan SL, Hou AJ, Huang JM (2009) Studies on the chemical constituents of *Coleus forskohlii*. Zhong Yao Cai 32(9):1381–1385

Wu H-Z, Yang Q-R, Yang Y-F, Liu Y-W (2007) Studies on HPLC-ELSD fingerprint of the *Coleus forskohlii* introduced in Tongcheng. Zhong yao cai = Zhongyaocai = J Chin Med Mater 30(11): 1370–1374

Wu M, Lai GF, Jin QD, Xu YL (2005) Spectral characteristic property of forskolins (2). Nat Prod Res Dev 17(Suppl):17–19. 22

Xu LL, Kong LY (2004) Isolation and identification of labdane diterpenoids from the roots of *Coleus forskohlii*. Chin J Nat Med 2:344–347

Xu LL, Kong LY (2006) Labdane diterpenoids from *Coleus forskohlii* (Willd.) Briq. J Integr Plant Biol 48:478–481

Xu LL, Lu J, Li WJ, Kong LY (2005) Studies on the chemical constituents in root of *Coleus forskohlii*. Zhongguo Zhong Yao Za Zhi 30:1753–1755

Xu YL, Jin QD, Liu J (2006) Spectral characteristics of forskolins (3). Nat Prod Res Dev 18 (Suppl):79–81. 97

Yanagihara H, Sakata R, Shoyama Y, Murakami H (1996) Rapid analysis of small samples containing forskolin using monoclonal antibodies. Planta Med 62:169–172

Yang QR, Wu HZ, Wang XM, Zou GA, Liu YW (2006) Three new diterpenoids from *Coleus forskohlii* Briq. J Asian Nat Prod Res 8:355–360

Yang WM, Jin QD, Xu YL (2007) Spectral characteristics of forskolins (4). Nat Prod Res Dev 19: 991–994

Yao CS, Shen YH, Xu YL (2002) The chemical constituents of *Coleus forskohlii*. Nat Prod Res Dev 14:1–6

Yao CS, Xu YL (2001) The diterpenoid quinones from *Coleus forskohlii*. Chin Chem Lett 12:339–342

Yuan H-R, Li M-G, Zhao J-Y, Wen M-L, Han X-L (2020) Recent advances in biosynthesis of forskolin. Zhongguo Zhong Yao Za Zhi 45(16):3790–3796. https://doi.org/10.19540/j.cnki.cjcmm.20200512.201. (Article in Chinese)

Zelnik R, Lavie D, Levy EC, Wang AHJ, Paul IC (1977) Barbatusin and cyclobutatusin, two novel diterpenoids from *Coleus barbatus* Bentham. Tetrahedron 33:1457–1467

Zerbe P, Hamberger B, Yuen MMS, Chiang A, Sandhu HK, Madilao LL, Nguyen A, Hamberger B, Bach SS, Bohlmann J (2013) Gene discovery of modular diterpene metabolism in nonmodel systems. Plant Physiol 162:1073–1091

Zhang W, Kong L (2009) Chemical constituents in the introduced *Coelus forskohlii*. Zhongguo Zhong Yao Za Zhi 34(16):2060–2062. (Article in Chinese)

Zhang XH, Zhang W, Jin QD, Xu YL (2005) Spectral characteristics of forskolins (1). Nat Prod Res Dev 17:163–165

Pharmacology of *Coleus forskohlii* and Forskolin

5

Abstract

Diterpene forskolin is the active constituent derived from the roots of *Coleus forskohlii*. It is endowed for its versatile pharmacological properties. Forskolin is the only plant-derived compound known to stimulate the enzyme adenylate cyclase and subsequently effecting cyclic AMP (cAMP) synthesis. Because of its broad range of activity in signal transduction reactions, the cAMP is known as the "second messenger." In the current chapter, some of the pharmacological properties including antiobesity, antidiabetic, antithrombotic, antioxidant, anti-inflammatory activity, asthma, glaucoma, heart disorders, hypertension, and anticancer activity of *C. forskohlii* and forskolin are discussed in detail.

Keywords

Coleus forskohlii · Forskolin · Antiobesity · Antidiabetic · Antithrombotic · Glaucoma · Asthma · Heart disorders · Anticancer

Abbreviations

AChE	Acetylcholinesterase
cAMP	cyclic adenosine monophosphate
IOP	Intraocular pressure
LDL	Low-density lipoprotein
LPS	Lipopolysaccharide
MBC	Minimum bactericidal concentrations
MFC	Minimal fungicidal concentration
MIC	Minimum Inhibitory concentration
SEM	Scanning electron microscopy
SOD	Superoxide dismutase

T. Pullaiah, *Forskolin*, https://doi.org/10.1007/978-981-19-6521-0_5

TG	Triglycerides
VLDL	Very low-density lipoprotein

5.1 Introduction

Geographic origin of the plant *Coleus forskohlii* is mostly from the subtropical regions of south Asia. Since ages, the root portion of the plant is used in Ayurvedic medicine. Typically the root of this plant is used as a marinated food, or pickle. This contains forskolin (7-β-acetoxy-8, 13-epoxy-1α, 6 β, 9α-trihydroxy-labd-14-ene-11-one), a diterpenoid. Forskolin is the only plant-derived compound known to stimulate the enzyme adenylate cyclase and subsequently effecting cAMP synthesis. Because of its broad range of activity in signal transduction reactions, the cAMP is known as the "second messenger." Forskolin as a rapid, and reversible activator of adenylyl cyclase underlies its wide range of pharmacological effects and is a valuable tool in studying the role of cAMP.

In cell lysates and intact tissues, forskolin activates adenylate cyclase probably by interacting with the catalytic subunit of the enzyme. Even at low concentrations, forskolin is reported to potentiate the hormonal activation and cAMP-dependent cellular responses. Thus, forskolin can be an invaluable tool for investigating the role of cyclic AMP in various physiological responses (Seamon and Daly 1981; Seamon et al. 1981).

A review article authored by Alasbahi and Melzig (2012) described the role of forskolin in modulating cAMP function. The secondary messenger, cAMP is involved in various cellular processes of the body, activates and inhibits disease pathogenesis as well as impacts the mechanism of action of many drugs. It also helps in attenuating the development or progression of fibrosis in the heart and lungs, regulating the coronary microvascular nitric oxide production, augmenting the myo-protective effects of ischemic precondition, stimulating injured retinal ganglion cells. Along with this, cAMP is also reported to show protective effect on exposure to ethanol, behavioral sensitization toward morphine and cocaine and cisplatin-induced oxidative injuries.

Forskolin is also reported to inhibit a number of membrane transport proteins and channel proteins via cAMP-independent pathway. Similar topographies are predicted on the membrane of many channel proteins and forskolin is predicted to bind these homologous sites (Laurenza et al. 1989). Forskolin can have a higher affinity to glucose transporter because of its structural homology with hexose molecule (Joost et al. 1988). It is also used to treat allergies, respiratory problems, cardiovascular diseases, glaucoma, psoriasis, hypothyroidism, weight loss, eczema, asthma, and hypertension (Patel 2010). And very recently, forskolin is used as a natural remedy to enhance the ability of antibiotics in treating urinary tract infections (UTI) (Abraham and Miao 2015).

5.2 Pharmacological Properties

Pharmacologically forskolin has been reported for many of its therapeutic properties since ages. In this concern, numerous preclinical and clinical studies have been carried out to prove its beneficial activities.

Seamon and Daly (1986), Valdes et al. (1987) and Alasbahi and Melzig (2010), Nisar et al. (2020) gave a detailed review on pharmacology of *C. forskohlii* and forskolin. Recently, Salehi et al. (2019) have discussed the therapeutic potential of forskolin. Many other reviews were published on the pharmacology of *C. forskohlii*, these include De Souza (1993), Reddy et al. (2005), Anonymous (2006), Wagh et al. (2009a), Kavitha et al. (2010), Patel (2010), Sharma and Vasundhara (2011), Soni and Singhai (2012), Kamini et al. (2013), Kaushal et al. (2013), Lakshmanan et al. (2013), Lakshmanan and Manikandan (2015), Saraswati et al. (2016), Bhowal and Mehta (2017), Lokesh et al. (2018), and Mitra et al. (2020). Sano (1983) reviewed the effect of forskolin on adenylate cyclase activity.

A summary of the effects of forskolin is shown in Fig. 5.1.

5.2.1 Antiobesity

Obesity, a complex metabolic disorder involves accumulation of excess body fat and thereby impairs the normal health of the individual. It causes adverse effects on BP, cholesterol, TGs, and insulin resistance and there by triggers the risk related to coronary heart disease, ischemic stroke, and type 2 diabetes mellitus. Along with this, 44% are prone to diabetes, 23% prone to ischemic heart disease, and 7–41% susceptible to cancer burdens. Unhealthy food habits and sedentary lifestyles are the major reason for the increased prevalence of obesity in children and adolescents

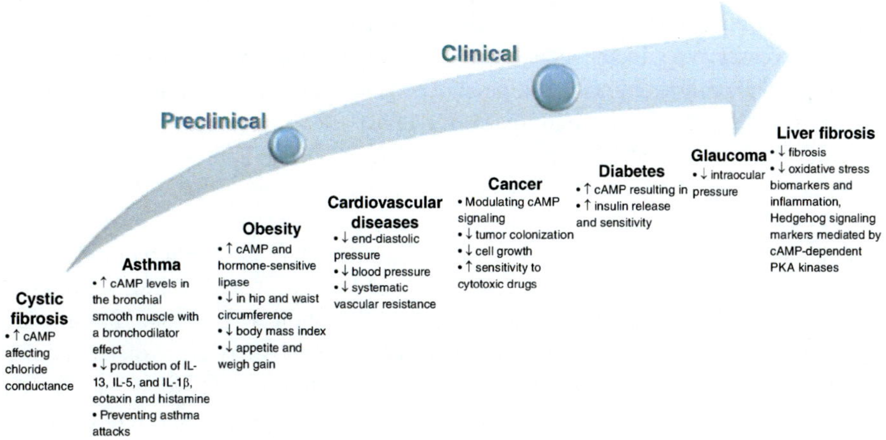

Fig. 5.1 Effects of Forskolin on human health (Source: Salehi et al. 2019)

(Shivaprasad et al. 2014). Majeed et al. (2021) reviewed the lesser investigated natural ingredients, including *C. forskohlii* in the management of obesity.

Despite a better understanding of the pathophysiology of obesity, only a few drugs are approved for use by regulatory authorities due to adverse effects (Ryan et al. 1999). Thus, in order to prevent the prevalence of obesity, many traditional medicines involving herbal sources are being used over thousands of years (Shivaprasad et al. 2014).

The Sami/Sabinsa group is a pioneer in providing the natural extracts for use in nutritional and cosmetic applications. One among them includes forskolin-enriched extract (ForsLean®) and an essential oil from the roots of *C. forskohlii*. But the water-insoluble feature of forskolin has limited its applicability. In this concern, a novel delivery system called Ocufors® has been developed to increase its usage and applicability in the management of glaucoma (Majeed 2012).

Administration of 50 mg/kg extracts of *C. forskohlii* showed reduction in body weight, food intake, and fat accumulation (Han et al. 2005). Increased food intake and weight gain in cafeteria diet-induced obesity in rats was significantly halted with the administration of *C. forskohlii* extract (CFE) (50 and 100 mg/kg). Normalization of increased levels of total cholesterol (TC), LDL, VLDL, TGs, and reduced HDL was observed with CFE treatment (Shivaprasad et al. 2014). A combinatory treatment of forskolin and rolipram for 10 weeks in female albino rats stimulated lipolysis and inhibited body weight increase by increasing cAMP levels. Compared to the individual treatment, combination therapy using the two agents was more effective in preventing diet-induced obesity (Doseyici et al. 2014). Lipolysis in fat cells is regulated by cAMP-dependent mechniasm (Schimmel 1984). Increased cAMP level due to activated adenylate cyclase is reported to subsequently phosphorylate and activate hormone-sensitive lipase (Belfrage et al. 1982). The relationship between cAMP production and lipolysis induced by forskolin in rat fat cells was investigated by Allen et al. (1986) and Okuda et al. (1992). The relationship between the cAMP-induced lipolysis varied with each treatment in the study. In the intact fat cells, forskolin-induced cAMP production and lipolysis were observed but at 10^{-5} M concentration lipolysis was observed with no increase in cAMP level. In other cases, i.e., in the cell homogenate, forskolin showed no stimulation of lipolysis and contrasting to the same, in a cell-free system consisting of endogenous lipid droplets and a lipoprotein lipase-free lipase fraction, forskolin stimulated both cAMP production and lipolysis prepared from fat cells. Thus, more experimental proofs are required to understand the relationship between cAMP and lipolysis (Okuda et al. 1992). Effects of forskolin and isoproterenol on the cAMP content of human adipocytes were investigated by Burns et al. (1987). Alterations in adipocyte cAMP concentrations in response to 100 μM forskolin and 10 μM isoproterenol were similar. Investigation by Litosch et al. (1982) indicate that forskolin increases production of cAMP in adipocytes through activation of adenylate cyclase. Lipolysis is activated by forskolin but at higher concentrations of total cAMP than catecholamines. Forskolin elevates cAMP and FFA release in rat adipocytes in a manner different from existing lipolytic factors (Ho and Shi 1982).

In order to manage the weight gain and decrease fat accumulation, forskolin is hypothesized in preclinical studies to enhance anabolic functions in the body to increase muscle mass. Further clinical study in 50 obese healthy subjects consuming forskolin for a period of 12 weeks showed enhancement in lean body mass, promoted fat loss, and improved the overall body composition (Majeed et al. 2009).

In an open field study, an extract of *C. forskohlii* root standardized for diterpene forskolin was tested for weight loss and lean body mass increase (Badmaev et al. 2002). The stimulated cAMP from diterpene of the plant is hypothesized to release fatty acids from the adipose tissue depots and thereby results in enhanced thermogenesis yielding low body fat and increased lean body mass (Girardier 1983). Six healthy overweight women tested for antiobesity effect by forskolin for 8 weeks showed a significant reduction in body weight and fat content, as well as promoted the lean body mass. Weight loss was statistically significant after 4 and 8 weeks with a mean change of 4.3 and 9.17 lbs., respectively. Also, at the maximum forskolin concentration of 50 mg/day there was no adverse effect on the systolic/diastolic blood pressure or pulse rate, thus indicating it as an effective weight management agent (Badmaev et al. 2002; Badmaev and Majeed 2004).

Similarly, a randomized study on 30 subjects consuming forskolin for 12 weeks showed favorable changes in body composition by significantly decreasing body fat percentage and fat mass. Also significant increase in bone mass, testosterone level was observed with the treatment (Michael et al. 2005).

Battochio et al. (2005) tested the effect of water extract of *C. forskohlii* (WEF) on weight gain, food energy utilization, and lipid metabolism in young rats with obstructive cholestasis. Cholestasis, independently of WEF, and WEF, independently of cholestasis both reduced ingested feed, energy utilization, and weight gain but there was no significant interaction between the two factors. Cholestasis, independently of WEF, increased liver wet weight, liver fat content, total serum cholesterol, and triacylglycerol. The WEF, independently of cholestasis, reduced these values, and there was a significant interaction between the two factors. The combination of Coleus extract with ephedrine and caffeine in doses of 20 mg and 200 mg, respectively, were found to be effective in the management of human obesity. Administration of these compounds separately to obese individuals was clinically ineffective (Badmaev and Majeed 2004).

An extract of *C. forskohlii* roots (ForsLean®) was tested by Tsuguyoshi (2004) in a 12-week open field study on overweight Japanese volunteers. This study indicates its usefulness in weight loss management with no apparent subjective and objective side effects (Tsuguyoshi 2004). Henderson et al. (2005), who evaluated the effectiveness of *C. forskohlii* (CF) on body composition, reported that CF has the potential to positively influence the loss and management of overall body weight.

The safety and efficacy assessment of ForsLean™ (250 mg of 10% CF extract) in a double-blind, randomized manner in 23 females for 12 weeks showed no adverse effects with the treatment. And significant differences were observed in caloric intake, body mass gain, scanned mass, and fat mass. The subjects treated with CF extract reported less fatigue, hunger, and also experienced stomach fullness (Krieder et al. 2005).

Kamohara (2016) reported that the extract of *C. forskohlii* had a substantial effect on weight loss and was safe and effective in reducing body fat in overweight/obese people. Chen et al. (2021) also reported that forskolin administration improved glucose metabolism and reduced fat cell diameter in high-fat diet-fed mice. Hayley (2015) also reported the antiobesity effect of *C. forskholii* extract. Tung et al. (2021) evaluated the antiobesity effects of *Garcinia indica* extract (GIE), *C. forskohlii* extract (CFE), and the combinations of these two extracts in 3 T3-L1 cells and high-fat diet (HFD)-induced obese mice. GIE, CFE, and the combinations of GIE and CFE were able to decrease body weight and adipocyte size by promoting fatty acid β-oxidation in HFD-induced obese mice.

C. forskohlii extract is used in weight loss products. The extract's alleged efficacy is attributed to forskolin. However, *C. forskohlii* extract (CFE) has been shown to induce fatty liver in mice, with components other than forskolin playing a part in this effect. Virgona et al. (2012, 2013) observed that CFE induced hepatic cytochrome P450 (CYP) while inducing fatty liver in mice, although these induction events were not seen with forskolin alone. The substance that induced fatty liver and hepatic CYP induction was soluble in ether and ethyl acetate (Yokotani et al. 2012). CYP was induced by CFE at a dose lower than that needed to induce fatty liver, but both phenomena seemed to be related. CFE-mediated CYP induction was clearly detected in a high-carbohydrate diet (Yokotani et al. 2013). Umegaki et al. (2014) reported that de novo synthesis and accumulation of triglyceride in the liver, is a major underlying mechanism of fatty liver induction by CFE.

5.2.2 Antidiabetic Effect

The elevation of cAMP levels by forskolin activates two signaling pathways: Protein kinase A (PKA) and guanine exchange by cAMP (Holz 2004). This results in a glucose-mediated response to pancreatic β-cells and insulin release (Ammon and Müller 1984). An in vivo study also supported the evidence of forskolin, causing a decrease in serum glucose levels, which decreased the severity of fasting hyperglycemia (Ríos-Silva et al. 2014). A clinical study on forskolin administration in conjunction with a hypocaloric diet in 41 patients revealed glucose-dependent insulin release and insulin sensitivity (Loftus et al. 2015). Moreover, forskolin has also shown attenuation of retinal inflammation in diabetic mice by limiting glucose transport into the retina (You et al. 2018).

Joost and Steinfelder (1987) reported that forskolin inhibits insulin-stimulated glucose transport in rat adipose cells by direct interaction with the glucose transporter rather than through activation of adenylate cyclase. Damera et al. (2019) reported that forskolin alleviates diabetic nephropathy via inhibition of aldose reductase and advanced glycation end product formation.

5.2.3 Antithrombotic Effect

Thrombosis are the blood clots in the veins caused due to injury, surgery, and certain medication. Clot formation includes two major principles including platelet aggregation and coagulation. Antithrombotic agents are used to prevent primary and secondary infection and to clear blood clots (Becker 2013). Forskolin as an antithrombotic agent inhibits platelet aggregation by stimulating adenylate cyclase activity, and augment prostaglandins' effects (Siegl et al. 1982; Adnot et al. 1982). It acts as a potential cerebral vasodilator. Forskolin fed at 10 μg/kg/min to rabbits showed an increase in blood flow in cerebrum, myocardium, and kidney, whereas a reduction in mean arterial pressure was noted (Wysham et al. 1986).

According to previous reports, the platelet aggregation induced by ADP, arachidonate, and collagen in human subjects, and in vivo models like rat and rabbit was potentially inhibited by forskolin treatment (de Souza et al. 1983; Agarwal et al. 1989). According to Agarwal et al. (1989) plasma adenosine plays an important role in forskolin's antiplatelet activity. Antiplatelet aggregation of forskolin elevates cAMP level. The amount of elevation correlates with the degree of ADP-induced aggregation inhibition (de Souza et al. 1983) and also relates to the progressive inhibition of fibrinogen binding in thrombin-stimulated human platelets (Graber and Hawiger 1982). Other than this, at low concentration, forskolin was reported to augment prostaglandins' and aspirin efficacy and potency (de Souza et al. 1983; Kariya et al. 1985; Salim 2003) in inhibiting platelet aggregation. Thus, both direct effect on adenylyl cyclase and receptor-mediated stimulation of the enzyme is attributed to forskolin (de Souza et al. 1983).

Christenson et al. (1995), who evaluated forskolin's effects on blood flow, platelet aggregation and metabolism, found that forskolin has a dose-dependent vaso-relaxing effect.

Synthetic vascular grafts have a thrombogenic surface and this impact leads to graft failure. The systemic pharmacologic interventions even though are used to lower platelet sequestration but these are associated with various side effects. Application of forskolin to inner surface of PTFE vascular grafts showed significant reduction in early platelet sequestration, as well as the graft potency was increased with the treatment (Christenson et al. 1989, 1991).

Beyond this, forskolin was found to exert an inhibitory effect on phospholipase C (de Chaffoy de Courcelles et al. 1987; Doni et al. 1988). The C-kinase activation plays a major role in enhancing the platelet's signal transduction. In this scenario, forskolin inhibits platelet secretion and aggregation by antagonizing C-kinase activity (de Chaffoy de Courcelles et al. 1987). Some of the other targets of the anti-aggregating activity of forskolin include increasing the nitric oxide synthase activity and inhibiting platelet-activating factor binding to the receptors (Russo et al. 2004; Wong et al. 1993).

5.2.4 Cardiovascular Effects

Forskolin was found to lower blood pressure and increase the heart's contractility. It works via vasorelaxation, causing relaxation of smooth muscles in the walls of blood vessels. This is believed to be due to forskolin's cAMP elevating ability, which relaxes the arteries and increases heart muscle contraction. It was also found to increase cerebral blood flow and enhance post-stroke recovery (Patel and Saraf 2016).

In isolated heart tissue, forskolin activates membrane-bound adenylate cyclase and a cytoplasmic cAMP-dependent protein kinase to a much higher degree than isoprenaline. Therefore, Metzer and Lindner (1981) postulated that the adenylate cyclase activation correlates with the positive inotropic effect via an enhanced calcium uptake by the heart muscle cell.

Bristow et al. (1983) reported a positive inotropic effect of forskolin. Forskolin was also found to exert positive inotropic actions on the isolated guinea pig heart, isolated rabbit heart, and on the dog and cat heart in situ (Lindner et al. 1978). In addition, forskolin augmented coronary blood flow in the isolated guinea pig heart. Forskolin increased the heart rate and lowered BP in dogs, cats, rats, rabbits, and spontaneously hypertensive and renal hypertensive rats (Bhowal and Mehta 2017). The cAMP increases the contractility by opening the slow Ca^{2+} channels, thus leading to elevation of intracellular calcium and the hypotensive effect by the increase of cAMP in the vascular smooth muscle (de Souza et al. 1983; Ammon and Müller 1985; Bhat et al. 1983; Lindner and Metzger 1983). However, the inotropic and chronotropic effects of forskolin in conscious dogs were reported to be due to mediated by neural mechanisms in addition to direct activation of adenylyl cyclase (Iwase et al. 1996). The cardiodynamic profile of forskolin was examined by Vaden and Adams (1985).

Abe and Karaki (1989) reported that forskolin exhibited a concentration-dependent inhibitory effect on vascular contractility of rat aorta by decreasing the cytosolic Ca^{2+} level at a lower concentration (0.1 μM) and decreasing the sensitivity of contractile elements to Ca^{2+} at a higher concentration (1.0 μM). According to Rembold and Chen (1998), repolarization and reduction in the intracellular Ca^{2+} sensitivity of force were found to be the primary mechanism of forskolin-induced relaxation of intact rat tail artery (White et al. 2000).

It has been shown by Bristow et al. (1984) that forskolin is a potent, powerful activator of myocardial adenylyl cyclase in human cardiac tissue preparations. In a clinical study by Ammon and Müller (1985), forskolin improved coronary blood flow and myocardial function without increasing the myocardial oxygen consumption. Moreover, in a comparative study, forskolin was found to induce a better cardiac performance than that produced by either dobutamine or by sodium nitroprusside in patients with stage III (NYHA) congestive cardiomyopathy (Baumann et al. 1990). Furthermore, forskolin was found to increase blood flow in the cerebrum and increase flow to the myocardium and kidneys despite a decrease in mean arterial pressure (Wysham et al. 1986).

Forskolin significantly lowers blood pressure via relaxation of vascular smooth muscle (Lindner et al. 1978; Dubey et al. 1981; Kramer et al. 1987; Schlepper et al. 1989). Intravenous forskolin administered at 3 μg/kg/min significantly reduced diastolic blood pressure (17%) and improved left ventricular function without increasing myocardial oxygen consumption (Kramer et al. 1987).

Forskolin at 4 μg/kg/min intravenously given to dilated cardiomyopathy patients resulted in decreased vascular resistance, increased heart rate, and improved left ventricle contractility (Schlepper et al. 1989). Lindner and Metzger (1983) reported that the action of forskolin on muscle cells is modified by hormones, Ca ions, and Ca antagonists.

A hypotensive effect was observed on i.v. administration of forskolin in anesthetized cats by Dohadwalla (1985). An i.v. dosage of 50 μg/kg forskolin for 16 min the duration of action was 16 min. The cardiovascular responses to bilateral carotid occlusion and vagal and preganglionic sympathetic nerve stimulation to challenges with acetylcholine, epinephrine, norepinephrine, and isoprenaline were not affected by 2 mg/kg of forskolin. This evidence suggests that the hypotensive action of forskolin may not be due to inhibition of central vasomotor tone or α or β adrenoreceptor inhibition (Majeed 2012).

The hypotensive effect of forskolin was studied in detail using normotensive anesthetized dogs (Dohadwalla 1985). Within 1–5 min of intravenous administration of 0.1 mg/kg of forskolin, a sharp fall in systolic and diastolic blood pressure of about 70 mm Hg was recorded. Simultaneously, the left ventricular pressure (dp/dt) was increased from 3100 to 4100 mm Hg/s, demonstrating its positive inotropic property.

The hypotensive effect of forskolin was further investigated in conscious spontaneously hypertensive (SH) rats (Dohadwalla 1985). Forskolin, in doses ranging from 2.5 to 10 mg/kg given intraperitoneally (i.p.) daily for 5 days, showed a dose-dependent fall in systolic blood pressure.

Studies by Dohadwalla (1985) suggest that forskolin's hypotensive action may be due to its direct smooth muscle relaxation property. The increase in cAMP levels in peripheral vascular smooth muscle cells mediates vasodilation and the ensuing hypotension. Jagtap et al. (2011) reported that *C. forskohlii* possesses antihypertension property in geriatric population.

5.2.5 Cystic Fibrosis

Cystic fibrosis is caused due to the mutation in the cystic fibrosis transmembrane conductance regulator (CFTR) gene (De Boeck and Amaral 2016). Two potential drug targets namely, potentiator VX-770 and corrector VX-809 linked to CFTR gene can be used to treat cystic fibrosis (Boj et al. 2017). CRE sequence (TGACaTCA) present in the promoter CFTR gene has thrown more light on cAMP regulation processes to gene expression (Matthews and McKnight 1996). Around 45–50% of cystic fibrosis patients suffer from a homozygous mutation named, F508DEL (Boj et al. 2017). In 1991, Drumm et al. reported that the association between CFTR and chloride conductance is sensitive to forskolin, where the order of sensitivity occurs at a similar level as the disease severity (Drumm et al. 1991). Several clinical studies, such as NCT03652090, NCT03390985, NCT03894657, and NCT02807415 are reported for cystic fibrosis, where forskolin is used to analyze drug sensitivity (Salehi et al. 2019).

5.2.6 Glaucoma

Several animal and human studies have demonstrated forskolin's ability to lower intraocular pressure (IOP), possibly via cAMP activation and a reduction in aqueous flow (Caprioli and Sears 1983; Badian et al. 1984; Caprioli et al. 1984; Seto et al. 1986; Meyer et al. 1987; Matsumoto et al. 1990).

Caprioli and Sears (1983) for the first time described the effect of forskolin on aqueous humor dynamics and intraocular pressure. The topical application of forskolin lowered the intraocular pressure. Effect of forskolin eye drops on intraocular pressure in healthy males was also reported by Badian et al. (1984). However, Lee et al. (1987) on monkeys and Brubaker et al. (1987) on humans reported that forskolin had no lasting effect on intraocular pressure in monkeys with glaucoma.

The ocular penetration of forskolin in suspension was investigated by Matsumoto et al. (1990). Only 0.03% of the instilled forskolin penetrated the ocular tissue. Topical 1% forskolin suspension caused cAMP increase in the aqueous humor 30 minutes after instillation, but cAMP returned to baseline level 60 minutes after instillation. The weak IOP lowering effect of topical forskolin suspension was considered to be due to its poor ocular penetration. However, a slight modification of molecular structure might increase ocular penetration. Authors suggested that only a slight increase in penetrative potential would be needed to make forskolin effective in antiglaucoma therapy.

Wagh et al. (2012) gave a detailed review on the antiglaucoma effects of forskolin. Wagh (2007) formulated forskolin ophthalmic drug delivery systems to test for its antiglaucoma efficacy. Ophthalmic Inserts (Tandale and Wagh 2011) of forskolin extract (OIE) and pure forskolin 98% (OIF) were prepared as matrix-controlled delivery to achieve once a day administration. Ophthalmic Insert Drug Delivery System (OIDDS) for forskolin showed a significant reduction in IOP up to 24 h and an increased corneal residence time up to 12 h with sustained therapeutic action (Wagh et al. 2008, 2009b, c; Wagh and Samanta 2010; Wagh and Pathare 2012).

Gupta and Samanta (2010) developed an ophthalmic delivery system of the forskolin, based on the concept of temperature-activated in situ gelations, which is likely to increase bioavailability. Poloxamer 407 (Pluronic F-127) is a block copolymer made of poly (oxyethylene) and poly (oxy propylene). An increase in temperature induces the sol–gel transition; however, it depends on the polymer's concentration and the presence of other additives. These formulations were therapeutically efficacious; reducing IOP for 12 hours and showed sol–gel phase transition (gelling) temperature of 22 °C and sustained drug release 72% in vitro behavior over a period of 4 h.

Dual-drug delivery system based on in situ gel-forming nanosuspension of forskolin to enhance antiglaucoma efficacy was developed by Gupta et al. (2010). In vitro experiments indicated that the formulation prolonged forskolin release for more than 5 h. The IOP was lowered by 31% in dexamethasone-induced glaucomatous rabbits and lasted for 12 h, which is significantly better than the effect of traditional eye suspension (18%, 4–6 h). Their investigations successfully proved that the pH and thermoreversible polymeric in situ gel-forming nanosuspension with controlled drug release ability exhibits a greater potential for glaucoma therapy.

Forskolin, a diterpene, is very poorly water soluble. Saettone et al. (2009) evaluated the solubilization of forskolin by some eye-compatible polymeric agents. While β- and 7-cyclodextrin were not particularly effective solubilizers, one polyoxyethylene-polyoxypropylene block copolymer (Pluronic^R F-127) increased the drug solubility by 40 times in water.

Majeed et al. (2014a, 2015b) assessed lowering of IOP using forskolin eye drops in the treatment of open-angle glaucoma. They used β-cyclodextrin to solubilize the molecule for ocular application. The trend towards a decrease in IOP pressure was higher in the forskolin group as compared to the Timolol group. Forskolin 1% w/v aqueous solution was found to be effective and safe in treating open-angle glaucoma. Forskolin 1% eye drops can be a safe alternative to beta blockers in glaucoma patients having concomitant asthma. The effect of forskolin 1% w/v aqueous solution is more effective than 0.5% Timolol eye drops. Forskolin Ophthalmic Solution developed by Sami-Sabinsa group (previously known as Sami Labs Limited) has been accorded approval in 2006. Subsequently, the efficacy of the Forskolin was observed at a much lower dose of 0.15% Forskolin solution (Majeed et al. 2014b, 2015b). Current treatment for glaucoma includes β-blockers and prostaglandin analogs, which have their own disadvantages.

Khan et al. (2018) reported that chitosan-coated poly lactic-co-glycolic acid (PLGA) nanoparticles amplify forskolin's ocular hypotensive effect. The particle size of optimized CS-PLGA NPs was found as 201.56 nm with a good PDI and positive zeta potential value with 72.32% entrapment efficiency and drug loading. Adenylate cyclase stimulation and ocular hypertension inhibition by forskolin analogs were also investigated by Yang et al. (2001).

Feng et al. (2016) isolated Isoforskolin (ISOF) from *C. forskohlii,* native to Yunnan in China. They found that *C. forskohlii* contained rich isoforskolin but not forskolin. ISOF was reported to activate adenylyl cyclase (AC) isoforms. The ophthalmologic and cardiovascular effects of ISOF were firstly reported in the 1990s. They found ISOF could lower blood pressure and intraocular pressure.

5.2.7 Asthma

Asthma is characterized by decreased cAMP levels in bronchial smooth muscle as well as high levels of possible adverse events (PAE). Mast cells, in response to allergenic stimuli, degranulate, releasing histamine causing bronchial smooth muscle contractions. Forskolin's activation of cAMP inhibits human basophil and mast cell degranulation, resulting in subsequent bronchodilation.

Forskolin relaxes guinea pig airways (Dohadwalla 1985). Kreuter et al. (1985) reported that forskolin blocked bronchospasms, the characteristic feature of asthma and bronchitis in guinea pigs caused by histamine and leukotriene. Forskolin produced a dose-dependent abolition of bronchospasms induced by histamine, acetylcholine, and serotonin in guinea pigs following intravenous administration. It is suggested that the bronchospasmolytic effect may also be brought about through stimulation of adenylate cyclase (Dohadwalla 1985). Bruka (1986) evaluated the bronchodilator activity of Forskolin for its potential use in the treatment of asthma.

Forskolin was tested double-blind and crossover in 12 healthy volunteers (nonsmokers) by whole-body plethysmography by Kaik and Witte (1986). The bronchodilating effect (after 5 min) was comparable to fenoterol at early time points, while fenoterol resulted in a stronger action at later time points.

A study involving humans revealed that inhaled forskolin powder formulations could cause bronchodilation in asthma patients (Bauer et al. 1993). Aerosolized dry forskolin powder resulted in a significant relaxation of bronchial muscles and relief of asthma symptoms (Bauer et al. 1993). In six asthmatics, nebulized forskolin administration resulted in an increase in forced expiratory volume. Two patients experienced immediate relief from bronchoconstriction and shortness of breath, while the remaining four patients reported relief after 10–15 min (Lichey et al. 1984).

Forskolin acts by increasing the bronchial smooth muscle's cAMP levels, which reduces bronchial reactivity and causes bronchodilator effect (Yousif and Thulesius 1999). Hiramatsu et al. (1994) explained the role of forskolin in tracheal muscles. Forskolin also possesses an activity that inhibits interleukin production (IL-13, IL-5, and IL-1β), eotaxin and histamine. Studies by Tsukawaki et al. (1987) have shown that forskolin relaxes airway smooth muscle in guinea pigs in vitro and in vivo by raising tissue cAMP levels and that its actions are independent of beta-adrenoceptors (Tsukawaki et al. 1987).

Kasonia (1995) screened methanol extract of six species including *C. forskohlii* (Syn.: *C. barbatus*). The extract showed relaxation activity on tracheal smooth muscle. González-Sánchez et al. (2006) determined the efficacy of forskolin in preventing asthma attacks. Isoforskolin significantly elevated cAMP levels in rat lung homogenate, and relaxed the histamine-induced contraction of isolated guinea pig trachea and lung smooth muscle (Feng et al. 2016).

5.2.8 Anticancer Activity

Many metastasizing tumor cell lines induce platelet aggregation both in vitro and in vivo. Upon aggregation, platelets release substances that promote tumor growth (Kim et al. 2004). Forskolin is a potent platelet aggregation inhibitor by stimulating platelet adenylate cyclase and increase of intracellular cAMP (Siegl et al. 1982).

Agarwal and Parks (1983) reported that forskolin at 2 μM inhibited the melanoma cell-induced human platelet aggregation. Administration of 82 μg of forskolin to mice 30–60 min prior to injection with a highly metastatic melanoma cell line (B16 F10) reduced tumor colonization in the lungs by 70% (Agarwal and Parks 1983).

Crude extracts of *C. forskohlii* roots, stems, and leaves were tested by Costa and Nascimento (2003) against pulmonary mucoepidermoid human carcinoma (cell NCI-H292) and epidermoid carcinoma of the larynx (cell HEp-2). Significant anticancer activity was observed against NCI-H292 cells by the acetonic and methanolic extracts of the roots. Sashidhara et al. (2007) isolated bioactive compound 13-epi-sclareol, which showed antiproliferative activity in breast and uterine cancers in vitro. Yamanaka et al. (2010) examined the expression and functional involvement of Hedgehog (Hh) signal transcription factors in pediatric tumor cells.

Huang et al. (2011) obtained seven compounds from ethyl acetate fraction of *C. forskohlii*. Their structures were identified as lupeol, oleanolic acid, uvalo, beta-sitosterol, colonic acid, demethylcryptojaponol, and coleolic acid. Compounds lupeol, oleanolic acid, demethylcryptojaponol, and coleolic acid showed antitumor activity. Moreover, compound lupeol is firstly found to have antitumor activity from the plants. Coleusin factor is a diterpenoid compound isolated from the root of *C. forskohlii*. Coleusin factor has been reported to suppress proliferation of and induce apoptosis in several types of cancer cells (Geng et al. 2007; Sun et al. 2011). Geng et al. (2014) reported that administration of Coleusin factor inhibits the growth of osteosarcoma xenografted in nude mice without systemic or immunologic toxicity.

Findings of Rajkumar and Malathi (2016) suggested that root ethanol extract of *C. forskohlii* possesses anticancer properties against gastric cancer cell lines through the apoptosis induction. Saeed et al. (2016) evaluated the cytotoxicity of the methanolic extract of *C. forskohlii* (Syn.: *Plectranthus barbatus*) leaves against two neoplastic leukemia lines, through resazurin assay. Although they correlated these results to the presence of forskolin and coleusin factor, they did not isolate them, but mentioned that these molecules showed cytotoxicity to murine melanoma cells (Agarwal and Parks 1983) and osteosarcoma (Geng et al. 2014). Borges et al. (2020) identified an antineoplastic activity of ethyl acetate fraction (EAF) of *C. forskohlii* leaves, but the identified constituents of EAF were plectrin, hydrolyzed abieatane, barbatusin, 3β-hydroxy-3-deoxibarbatusin, cyclobutatusin, luteolin, rosmarinic acid, and coleoside B.

Abd-Allah et al. (2017) investigated the hormonal and immunological responses to *C. forskohlii* treatment in female rats with experimentally polycystic ovaries syndrome. Opioid and immune systems were impaired in hyper androgenized rats (Polycystic Ovarian Syndrome model) and the *C. forskholii* treatment could restore most of these functions. Sapio et al. (2017) described some features of cAMP signaling that are relevant to cancer biology and address the state of the art concerning the natural cAMP elevating compound forskolin and its perspectives as an effective anticancer agent.

Illiano et al. (2017) reported that forskolin exerts significant growth and migratory inhibitory effects on Panc-1 and AsPC-1 pancreatic cancer cells. Forskolin strongly enhances gemcitabine-induced antiproliferative effects by both cell cycle inhibition and cell death induction. Illiano et al. (2018) investigated forskolin's effects on the sensitivity of MDA-MB-231 and MDA-MB-468 TNBC cells to doxorubicin through MTT assay, and cell cycle progression and cell death by flow cytometry, cell number counting, and immunoblotting experiments. They demonstrated that forskolin strongly enhances doxorubicin-induced antiproliferative effects by cell death induction.

Ganash and Qanash (2018) reported that *C. forskohlii* has a promising effect on tumor activities by inhibiting cell proliferation with the IC_{50} 89.13. Hayashi et al. (2018) reported that forskolin increases the effect of everolimus, an anticancer drug, on aromatase inhibitor-resistant breast cancer cells. De Freitas et al. (2018) extracted proteins from *C. forskohlii* (Syn.: *Plectranthus barbatus*) and evaluated their cytotoxicity against A549 and RAW264.7 cancer cells. The aqueous extract from the leaves presented cytotoxic action against lung carcinoma tumor cell lines.

Mothana et al. (2019) investigated the cytotoxic activities of crude extract of *C. forskohlii* and its isolated compounds. The crude ethanolic extracts of the *C. forskohlii* showed remarkable cytotoxic activity against Hela, HepG2, and HT-29 cancer cell lines with IC_{50} values of 10.1, 10.7, and 32.0 μg/mL, respectively, which were comparable with the positive control values. Four compounds (Coleonol B, forskolin, sugiol, and 5,6-dehydrosugiol) were isolated from chloroform fraction and ferruginol from hexane fraction. The isolated diterpenoids showed interesting cytotoxic effects with IC_{50} values between 15.1 and 242 mg/mL. The greatest activity was shown by ferruginol, followed by 5,6-dehydrosugiol and forskolin. Sugiol and Coleonol B exhibited a moderate to weak cytotoxic activity (Mothana et al. 2019). Coleon C extracted from Coleus is useful as inhibitor for tumor growth and tumor cell proliferation (Liu et al. 2007).

Wang et al. (2019) reported that forskolin exerts anticancer roles in non-Hodgkin's lymphomas (NHL) via regulating Axin/β-catenin signaling pathway. Moreover, forskolin improves the effects of SP600125 on cell apoptosis enhancement and tumorigenesis inhibition of NHL cells.

Cordeiro et al. (2022) evaluated the antineoplastic activity of *C. forskohlii* (Syn.: *Plectranthus barbatus*) leaf aqueous extract (CFA) and acetone: water (7:3) organic extract (CFO). Cytotoxicity in normal peripheral blood mononuclear cells at concentrations ranged between 0.1 and 100 μg/mL and in neoplastic cell lines Toledo, K562, DU-145, and PANC-1 at 1, 10, and 100 μg/mL.

5.2.9 Antioxidant Activity

A comparative assay was made between crude ethanolic extract of *Peumus boldus* and *C. foskohlii* in order to evaluate their relative in vitro antioxidant capacities by Tamasiro et al. (1998). The concentration necessary to decrease 50% of the spontaneous autoxidation of the system indicates that *C. forskohlii* (29 μg/mL) was almost 10 times less active than *P. boldus* (3.2 μg/mL).

In comparison to other *Coleus* species, a higher amount of polyphenols and higher antioxidant activity is reported in the extract of *C. forskohlii*. According to Rasineni et al. (2008), the leaf extract of *C. forskohlii* contains a significantly higher amount of total polyphenols, flavonols, and flavones and also exhibit high antioxidant activity. The plant is also a rich source of diterpenoids with different oxygen patterns. Till 2001, approximately, six diterpenoids were isolated from whole plant. In the same year, Yao and Xu (2001) have isolated two new diterpenoidquinones and named them as coleon S and T.

Maioli et al. (2010) assessed the antioxidant activity of the water extract of *C. forskohlii* leaves on Fe^{2+} $^{-}$citrate-mediated membrane lipid peroxidation in isolated rat liver mitochondria and in non-mitochondrial systems. Treatment at 15–75 μg/ml *C. forskohlii* extract showed significant reduction in DPPH, OH scavenging activity, and iron chelation. Along with this, mitochondria protection against Fe^{2+}/citrate-mediated swelling and malondialdehyde production was also reported. According to authors, the extract containing nepetoidin—caffeic acid esters as phenolic compounds is probably involved in all the pharmacological changes and thus accounts for hepatoprotective action.

The antioxidant status of various parts of *C. forskohlii* was analyzed by Khatun et al. (2011). The enzymatic antioxidant properties measured in terms of the activities were significantly higher in tubers than in the leaves, roots, and stems. Besides their medicinal properties, the tubers possessed significantly rich sources of both enzymatic and nonenzymatic antioxidants.

Ganash and Qanash (2018) reported that *C. forskohlii* has different antioxidant enzymes, including catalase, polyphenol oxidase, and peroxidase. Recently, Mothana et al. (2019) evaluated the antioxidant activity of crude extracts of *C. forskohlii* and its isolated compounds. *C. forskohlii* extract showed the ability to inhibit β-carotene's discoloration at a 1000 μg/mL concentration with a total antioxidant value of 66%. Results of the DPPH radical scavenging method demonstrated comparable free radical scavenging activity. Also, the chloroform fraction of *C. forskohlii* exhibited the highest antioxidant and free radical scavenging activity among the tested fractions with values of 76 and 80% at 1000 μg/mL. Among the isolated compounds, sugiol, 5,6-dehydrosugiol and ferruginol were able to inhibit the discoloration of β-carotene with total antioxidant values of 65%, 67%, 69%, respectively. Ibrahim et al. (2018) investigated the antioxidant activity of plant and callus cultures of *C. forskohlii* (Syn.: *Plectranthus barbatus*). Aqueous methanol extracts exhibited higher DPPH radical scavenging activity than hexane extracts at all tested concentrations.

Essential oils (EOs) from the roots, stems, and leaves of *C. forskohlii*, cultivated in northern Italy, were obtained by steam distillation. The highest yields were obtained from roots (268.15), followed by leaves (64.34 mg/kg), and stems (19.76 mg/kg). A total of 128 structures were identified. Fe^{++} chelating and antiradical activities (DPPH and ABTS) were evaluated: root and stem EOs showed the strongest activities, while EOs from leaves did not show relevant activities (Gelmini et al. 2015). Shanmugam et al. (2018) reported antioxidant activity of rhizome extracts of *C. forskohlii*. Total phenol and flavonoid contents were found to be, respectively, 38.82Â ± 0.22 mg gallic acid equivalents/g and 21.34Â ± 0.32 quercetin equivalent/g.

Fadlelmula et al. (2020) investigated the antioxidant activity and total phenolic content of *C. forskohlii.* Butanolic extract showed the highest radical scavenging followed by ethyl acetate extract. Where the *n*-butanol exhibited highest Total Phenolic Content (TPC) value, followed by ethyl acetate then petroleum ether, ethanol, and eventually chloroform extract exhibited the lowest (TPC) value.

5.2.10 Anti-Inflammatory Activity

The hexane, chloroform, methanol, 80% methanol, and water extracts of *C. forskohlii* were screened for in vitro anti-inflammatory activity. Methanolic and water extracts showed maximum activity (James 2003). Similarly, extracts from 12 medicinal plants including *C. forskohlii* of Kenya showed anti-inflammatory activity (Matu and van Staden 2003). Intraperitoneal administration of forskolin to rats showed significant inhibition of Carrageenan-induced paw edema in a dose-dependent manner (Rupp et al. 1986).

Menon and Latha (2011) screened the extracts of *C. forskohlii* for its in vitro anti-inflammatory activity. The antioxidant and anti-inflammatory effect accessed by using DPPH and BSA anti-denaturation and HRBC membrane stabilization assay indicated that among the extracts, water and methanolic extracts showed maximum anti-inflammatory property. Metabolic profiling of the methanolic extract by using thin layer chromatography confirmed the presence of forskolin as a major bioactive compound.

Forskolin was evaluated at a dose of 10 mg/kg body mass for its anti-inflammatory and anti-nociception properties in rats in comparison with diclofenac by Suresh et al. (2018). The percent inhibition in inflammation exhibited, by diclofenac were found to be 0, 18, 37, 20, and 17.2%, and by forskolin 0%, 11%, 40%, 31%, and 6.8% at 0, 30, 60, 90, 120 min drug reaction time, respectively. Diclofenac increased the tail flick latency time at 30, 60, and 90 min forskolin at 90 min.

Xiao et al. (2021) reported that isoforskolin alleviates AECOPD by improving pulmonary function and attenuating inflammation which involves downregulation of Th17/IL-17A and NF-κB/NLRP3.

5.2.11 Antidepressant Effect

The imbalance level of neurostimulators such as serotonin and dopamine in brain is associated with depression. In order to address the issue, supplements involving 5-HTP, tryptophan, or SSRI drugs like Prozac or Zoloft are used. In case of catecholamine neurotransmitters like epinephrine, and norepinephrine deficiency, the amino acids L-Phenylalanine or L-Tyrosine, or monoamine oxidase inhibitors like GeroVital (GH3) or Deprenyl are reported to be helpful. As per recent research, drugs that elevate the catecholamines by increasing cAMP levels can be better agents to improve neurostimulatory function. In accordance with this, forskolin at 0.01–0.1 mg/kg showed strong antidepressant activity by significantly decreasing immobility ratings. The minimum dosage of 0.01 mg/kg dose showed 150 times more potency than the standard drug (amitriptyline-15 mg/kg) (Maeda et al. 1997). In a clinical study, forskolin infusion in a 75-min duration to four depressed and five schizophrenic patients showed transient mood elevation or stimulation (Bersudsky et al. 1996).

5.2.12 Antistress Effect

Tiwari et al. (2014) investigated the antistress activity of forskolin. Elevated plus maze (EPM) model and forced swimming test (FST) revealed that forskolin has antistress activity.

5.2.13 Antidyspeptic Activity

The aqueous extract of *C. forskholii* is reported to induce antidyspeptic activity by decreasing gastric secretion and it also shows protection against stress-induced gastric ulcers (Lygia et al. 1991).

5.2.14 Anti-Ulcer Activity

Plectrinon A one of the active ingredients of *C. forskohlii* inhibits the gastric H^+/K^+-ATPase, more effectively than the classic proton pump inhibitor omeprazole. The inhibition of these proton pumps in the stomach effectively helps in regulating ulcer formation (Li et al. 2006).

5.2.15 Hepatoprotective Activity

Forskolin and 1,9-dideoxyforskolin are efficacious agonists of the pregnane X receptor (PXR, NR1I2) and thus there is an activation of PXR which mediates the hepatoprotective effect (Jeff et al. 2006).

Staudinger et al. (2006) reviewed in detail the hepatoprotective activity of forskolin and *C. forskohlii* extracts. Forskolin and 1,9-dideoxyforskolin induce CYP3A gene expression in primary cultures of rodent hepatocytes (Sidhu and Omiecinski 1996). An approach has been made to evaluate the effect of *C. forskohlii* and its major constituents on cytochrome P450 (CYP3A, CYP2B, and CYP2C) mRNA expression in rat hepatocytes by Nagarajappa et al. (2016). The test substances did not show any significant mRNA expression compared to the control against CYP3A, CYP2B, and CYP2C. It can be concluded from this study that *C. forskohlii* and its major constituents may not be involved in CYP450 induction-based drug interaction. Recent studies have revealed that both forskolin and 1,9-dideoxyforskolin are efficacious PXR agonists (Ding and Staudinger 2005; Dowless et al. 2005). The findings suggest that herbal therapy with *C. forskohlii* extract should be approached cautiously in patients on combination therapy, due to the potential for herb–drug interactions (Staudinger et al. 2006). El-Agroudy et al. (2016) reported that forskolin significantly reduced hepatic fibrosis induced by CCl_4.

5.2.16 Antimicrobial Activity

Ethanolic extract of the *C. forskohlii* roots was shown to exhibit marked inhibitory action against *E. coli* toxin-induced secretory response in rabbits and guinea pig ileal loops (Yadava et al. 1995). Water, hexane, and methanol extracts of 12 plant species were screened for in vitro antibacterial activities by Matu and van Staden (2003). The highest activity was found in the methanol extracts of *C. forskohlii* against *Staphylococcus aureus*.

Nilani et al. (2006) evaluated the antifungal activity of solvent extracts of *C. forskohlii* against *Aspergillus niger, A. fumigatus, A. ruantii, Proteus vulgaris,* and *Candida albicans*. The petroleum ether extract of *C. forskohlii* exhibited significant antifungal activity against all the selected organisms. Similar activity was observed against *C. albicans* (ATCC 90028) by Runyoro et al. (2006).

Bodiwala et al. (2009) evaluated various extracts of the aerial parts of *C. forskohlii* at their noncytotoxic concentration against HIV-1 NL4–3. Chloroform, ethyl acetate, and n-butanol extracts showed 45.6, 66.5, and 37.7% inhibition of HIV, respectively, in CEM-GFP cells infected with HIV-1NL4–3 at 5 μg/mL. Four diterpenes, 1-deoxyforskolin, 1,9-dideoxyforskolin, forskolin, and isoforskolin were isolated from the chloroform extract and tested against the virus. 1-Deoxyforskolin and forskolin were found to be active against HIVNL4–3.

Aqueous extract of *C. forskohlii* is traditionally used as an anti-inflammatory and antifungal agent. The effect of this extract and of its main component, rosmarinic acid, on the viability of the cariogenic bacteria, *Streptococcus sobrinus* and *S. mutans*, was determined by MIC and MBC by Figueiredo et al. (2010). The influence of this extract on the biofilm formation and the inhibition of glucosyltransferase enzyme produced by this species were also analyzed. The aqueous extract of *C. forskohlii* was a stronger inhibitor than rosmarinic acid with MIC values of 3.8 mg/mL for *S. sobrinus* and 2.9 for *S. mutans* compared to 8.4 and 7.3 mg/mL for rosmarinic acid. Similarly, *C. forskohlii* showed better inhibition of biofilm formation and glucosyltransferase inhibition activity, suggesting that the extract may be useful in the prevention of dental caries (Figueiredo et al. 2010).

Senthilkumar et al. (2010) evaluated the antibacterial activity of *C. forskohlii* against five human pathogenic bacteria *Staphylococcus aureus*, *S. epidermidis, Salmonella typhi, Klebsiella pneumoniae*, and *Vibrio parahemolyticus*. Their study revealed a remarkable antibacterial activity against Gram-negative bacterial strains than Gram-positive bacterial strains. The most effective activity was shown by

C. forskohlii with MIC ranging from 15 mm against *S. typhi* and 14 mm with *S. aureus*. The extracts inhibited *K. pneumoniae* and *V. parahemolyticus* with 8 mm and 10 mm, respectively.

Manikandan et al. (2018) showed that silver nanoparticles of *C. forskohlii* root extracts exhibit antibacterial activity against two human bacterial pathogens such as *Bacillus subtilis* and *Alcaligenes faecalis*. Saklani et al. (2011) evaluated the antimicrobial activity of the extracts (Pet. ether, diethyl ether, chloroform, methanol, and ethanol) of roots *C. forskohlii* against bacteria *Staphylococcus aureus, Pseudomonas fluorescens, Sericea, Klebsiella pneumoniae, Bacillus pumilus* and fungi *Aspergillus flavus, A. parasiticus, Trichoderma rubrum, Microsporum gypseum*. The different extracts of the plant *C. forskohlii* were found to have maximum antibacterial and antifungal activity.

In vitro antibacterial activity of crude root extract of *C. forskohlii* was investigated by Anbuselvan and Muralikrishnan (2013) against *Escherichia coli, Pseudomonas aeruginosa, Salmonella typhi, Klebsiella pneumoniae, Staphylococcus aureus,* and *S. epidermidis*. Gram-negative bacterial strains were more susceptible to the crude extracts as compared to Gram-positive bacterial strains. However, this study revealed maximum growth inhibition and effectiveness were remarkably observed in the extracts of *C. forskohlii*.

The antimicrobial activity of different extracts of *C. forskohlii* roots was reported by Baskaran et al. (2011) against various bacterial and fungal species. In vitro antibacterial activity of crude extracts of roots, shoots, and leaves of *C. forskohlii* were screened by Malleswari et al. (2013) against *Bacillus subtilis, Pseudomonas fluorescens, P. aeruginosa, K. pneumoniae, E. coli, S. aureus,* and *Streptococcus pneumoniae*. Gram-negative bacterial strains were more susceptible to the crude extracts as compared to Gram-positive bacterial strains.

Mwitari et al. (2013) reported that *C. forskohlii* (Syn.: *Plectranthus barbatus*) has both bactericidal and fungicidal activities. Figueiredo et al. (2014) evaluated methanolic extract of leaves from *Plectranthus barbatus* against oral pathogens *Streptococcus mutans* and *S. sobrinus*. The extract showed bacteriostatic activity with MIC of 0.3 mg/mL. The MBC obtained for extract was 0.6 mg/mL against *S. sobrinus* and 0.8 mg/ml against *S. mutans*. When methanol extract of *P. barbatus* was used to inhibit the growth of the two bacterial strains in biofilm, IC_{50} value was 1.9 mg/mL against *S. sobrinus* biofilm and 0.7 mg/mL against *S. mutans* biofilm, respectively. *P. barbatus* IC_{50} values for the biofilm formation were 0.63 mg/mL and 0.13 mg/mL against *S. sobrinus* and *S. mutans,* respectively. GTF from *S. sobrinus* was inhibited at 30% when 0.3 mg/ml of *P. barbatus* extract was used. Veríssimo et al. (2013) also reported antimicrobial activity of *C. forskohlii* (Syn.: *Plectranthus barbatus*) extract against *Staphylococcus aureus, S. epidermidis* (6.25 mg/mL), *Streptococcus pneumoniae* (6.25 mg/mL), and *Escherichia coli*.

Kala (2014) evaluated the effect of petroleum ether, methanol, ethyl acetate, acetone, and chloroform extracts of *C. forskohlii* on Gram-positive bacteria like *Bacillus subtilis, Staphylococcus aureus* and Gram-negative bacteria *E. coli, Pseudomonas aeruginosa* and the yeast *Candida utilis*. All the extracts showed significant activity against all pathogens.

Sasikala et al. (2014) evaluated the antibacterial activities of *C. forskohlii* in different extracts at 80 μl against three bacteria, namely *Bacillus cereus, Escherichia coli*, and *Lactobacillus*. The water and acetone extracts exhibited the highest antimicrobial activity in almost all the pathogenic bacteria. Chloroform root extract on *E. coli* showed MIC 26.33 while root water extract showed MIC 25.5. Acetone root extract showed MIC 22.83 on *B. cereus* while petroleum root extract showed MIC 22.0 on *Lactobacillus*. The water extracts were much active against all the bacteria tested.

A study on antimicrobial efficacy of *C. forskohlii* against *Staphylococcus aureus* showed both bacteriostatic and bacteriocidal activity at MIC values ranging from 60 to 300 μg/ml (Snowden et al. 2014). Nidiry et al. (2015) showed that forskolin is one of the antifungal compounds present in the *C. forskohlii*, where root extracts exhibited mycelia growth inhibition of *Colletotrichum gloesporoides* and spore germination inhibition of *Alternaria solani*.

Rajkumar and Malathi (2015) investigated the antimicrobial activity of *C. forskohlii*. Ethanolic extract of *C. froshkolii* roots showed highest antibacterial activity compared to stem and leaf. The highest antimicrobial activity was observed against *Klebsiella pneumoniae* (19 mm) and *Candida albicans* (16 mm) in ethanolic extract of root.

The antimicrobial activities of various extracts of *C. forskohlii* roots were studied by Singh and Singh (2016). The ethanolic extract was more effective against *Bacillus cereus, Micrococcus luteus, Klebsiella pneumoniae*, and *Staphylococcus aureus*, whereas acetone extract was more effective against *Micrococcus luteus, Bacillus cereus*, and *Klebsiella pneumoniae*. The different extracts were also found to be effective against the fungi such as *Aspergillus niger, A. flavus, Candida albicans, C. tropicalis*, and *Cryptococcus neoformans*.

C. forskohlii extracts in different solvents were tested against some gastrointestinal pathogens by Mathur et al. (2011). In the case of *Escherichia coli*, the inhibition was recorded by the treatment of all the extracts. *Staphylococcus aureus* was found to be resistant to hexane extract while ethanol extract inhibited *S. aureus*' growth. In the case of *Salmonella typhimurium*, all extracts showed good inhibitory effects. In the case of *Staphylococcus epidermidis* compared to control, water extract showed more inhibition than other extracts.

Atulkar et al. (2015) evaluated the antimicrobial activity of ethanolic extract of *C. forskohlii* root against *Staphylococcus aureus, Pseudomonas aeruginosa, E. coli, Bacillus subtilis*, and *Aspergillus niger*. *S. aureus* is the most sensitive strain (12.03 mm) of those tested while other microorganisms showed less sensitivity.

Ganash and Qanash (2018) reported that *C. forskohlii* has antibacterial activities against *Staphylococcus aureus, E. coli*, and *S. typhi* while weak antifungal activity was observed. Mothana et al. (2019) evaluated the antimicrobial activity of crude extracts and isolated compounds of *C. forskohlii*. MIC, MBC, and MFC of the extracts and isolated compounds were investigated. The most sensitive strains were the Gram-positive bacteria *Streptococcus mutans* and the fungal strain *Cryptococcus neoformans*, which were more susceptible to *C. forskohlii* extracts with MIC values between 62.5 and 250 mg/mL than the other microbial strains.

Nguta and Kiraithe (2019) reported antimicrobial activity of *C. forskohlii* methicillin-resistant *Staphylococcus aureus*, *Escherichia coli, S. aureus*, and *Bacillus cereus.* The crude ethanolic rhizome extract was tested by Shanmugam and Pradeep (2019) in three different concentrations (10, 20, and 30 mg/mL) for antibacterial activity. Maximum antibacterial activity was noted against *Bacillus subtilis* (15 mm), followed by *B. cereus* (14 mm), *Pseudomonas aeruginosa* (12 mm), and *Staphylococcus aureus* (12 mm). *E. coli* was not inhibited even at the highest concentration tested i.e. 30 mg/ml.

Essential oils (EOs) from the roots, stems, and leaves of *C. forskohlii*, cultivated in north Italy, were evaluated for their in vitro antimicrobial activity, showing optimal growth-inhibition in antibiogram ($\varnothing > 35$ mm) and MIC tests (32–64 μg/mL) against *Candida albicans*, while EOs from leaves showed a good activity ($25 < \varnothing < 34$ mm, MIC 64–128 μg/mL) against *Escherichia coli* (Gelmini et al. 2015). Ramkumar et al. (2019) reported antimicrobial activity of rhizome extract of *C. forskohlii* (Syn.: *Plectranthus barbatus*) against *Escherichia coli, Staphylococcus aureus*, and *Bacillus subtilis* species of bacteria and *Candida albicans* and *Aspergillus flavus* species of fungi.

Fadlelmula et al. (2020) evaluated the antimicrobial activity of *C. forskohlii* from Saudi Arabia, where *n*-butanol showed the maximum inhibition. Khatun (2020) studied the antimicrobial activity of tuber extracts of *C. forskohlii*. Higher inhibition zone was recorded in *Pseudomonas* sp. which was 3.5 cm. *S. aureus* showed lowest inhibition zone of 2.0 cm. Higher mycelial growth inhibition was found in *Curvularia lunata* (76.70% at 11 μg/mL tuber extract) followed by *Alternaria alternata* (68.20% at 11 μg/mL tuber extract) and *Fusarium oxysporum* (65.84% at 11 μg/mL tuber extract).

Al-Ghamdi (2021) investigated the antimicrobial activity of *C. forskohlii* stem extract. The antimicrobial activity showed that all microorganisms tested were resistant at the concentration of 25 and 50 mg/ml of plant extracts, whereas the concentrations of 100, 150, and 200 mg/mL showed varying activities. Alghamdi et al. (2021) investigated the bactericidal activity of *C. forskohlii* extract against multidrug-resistant *Acinetobacter baumannii* strains isolated from hospital. The MIC of the extract against the pathogenic strains was approximately 4 mg. The plant extract has the ability to kill MDR *Acinetobacter* strains.

Chakraborty et al. (2022) evaluated the antimicrobial activities of in vitro and ex vitro grown plants of *C. forskohlii* against 20 MDR strains of urinary tract infections (UTI) bacteria. When methanolic leaf extracts were tested, 17 among 20 UTI pathogens were found to be sensitive at only 0.75 mg/cup concentration, and three were resistant even at the 6.0 mg/cup concentration. Both the in vitro and ex vitro plants exhibited the highest efficacy against the pathogenic strain of *Klebsiella* with the ZI 17.66 ± 0.33 and 18.66 ± 0.57 mm, respectively.

Shaker et al. (2022) investigated the antimicrobial activities of *C. forskohlii* fractions against Gram-positive and Gram-negative bacteria and fungus *Candida albicans* and identified major active compounds. The antimicrobial assays revealed that ethyl acetate was the most potent fraction and the major abundant metabolites of

C. forskohlii. Thymoquinol-2-*O*-*β*-glucopyranoside, syringic acid, methyl 3,4,5-trihydroxybenzoate, and luteolin were isolated herein for the first time.

Cordeiro et al. (2022) evaluated the antimicrobial activity of *C. forskohlii* leaf aqueous extract (CFA) and acetone: water (7:3) organic extract (CFO). CFO showed bacteriostatic activity against *Acinetobacter baumannii* (MIC = 250 μg/mL) clinical isolate and CFA fungistatic activity against *Trichophyton rubrum* (MIC = 800 μg/mL).

5.2.17 Hypothyroidism

Forskolin is reported to increase thyroid hormone production and stimulate its release by activating guanine nucleotide-binding proteins (Saunier et al. 1990). The stimulation of thyroid hormone can in turn promote normal body weight and thereby contributes to the antidepressant effects.

Laurberg (1984) compared the effects of 10^{-5} M forskolin and 100 mu units/ml TSH on the dynamics of T4 and T3 secretion from perfused dog thyroid lobes. Both agents induced pronounced increases in T4 and T3 secretions. The increase in secretion was significantly steeper during forskolin than during TSH stimulation.

5.2.18 Skin Problems

According to Majeed and Prakash (2007), *Coleus* oil can be used as a useful, topical preparation due to its antimicrobial properties. The ingredient has shown effective inhibition of *Propionibacterium acnes*, and other microorganisms responsible for skin infections and eruptions.

Psoriasis is an immune system problem affecting the skin cells to form scales, itchy and dry patches. It can be alleviated by normalizing the cAMP/cGMP ratio. Similar to asthma, decreased levels of cAMP and cGMP levels are reported in psoriasis. Supplementation of forskolin to psoriatic patients in a randomized clinical trial showed regulation of the monophosphates level and thereby improvising their condition (Ammon and Müller 1985).

5.2.19 Antispasmodic Effect

Based on the screening of Indian plants for their effective biological activity, the extract of *C. forskohlii* roots effectively lowered the blood pressure and also showed antispasmodic effect (Dubey et al. 1974). The contractility of the guinea-pig ileum with the treatment of *C. forskohlii* essential oil (PBEO) at concentrations ranging from 1 to 300 μg/mL was studied. The oil was reported to contain α-pinene, myrcene, and caryophyllene as the major ingredients. In comparison with the standard drugs like alpha-pinene and papaverine, treatment with PBEO decreased the basal tonus of the ileum with a maximal response of 62.7 ± 3.8%. PBEO also blocked the phasic contractions evoked by acetylcholine as well as decreased the contractions induced by histamine or barium chloride. In addition, PBEO relaxed the precontracted tissues and at the higher concentration (300 μg/mL) it decreased the maximal response of calcium chloride-induced contraction in depolarized tissues. These results suggest PBEO as an effective intestinal relaxant and antispasmodic ingredient (Câmara et al. 2003).

5.2.20 Protection against Gastric Ulcers

Painful inflammation of stomach lining due to bacterial infection and excess usage of anti-inflammatory drugs causes gastric ulcers. The water extract of stems and leaves of *C. forskohlii* given at 1–10 g/kg, p.o. to rats and mice was checked for its protective effect against gastric ulcers. After the treatment, increase in the intestinal transit by 30% and reduced gastric secretion was observed at 2 g/kg of the *C. forskohlii* extract. Also, reduction in total acid secretion and increase in gastric pH from 34.4 ± 11.0 to 2.7 ± 0.5 mEq/L and 2.2 ± 0.3 to 6.5 ± 0.8, respectively, were reported. Thus, according to the study, the water extract of *C. forskohlii* potentially reduces the gastric secretion and thereby protects the stomach lining from gastric ulcers (Fischman et al. 1991). Similar results were reported by Schultz et al. (2007). The water extract of *C. forskohlii* leaves injected at 0.5–0.1 g/kg into the duodenal lumen decreased the volume and total acidity of gastric acid secretion in pylorus-ligated mice.

5.2.21 Bone Formation

An in vivo study, to evaluate the effect of forskolin in the bone formation was reported by Doorn et al. (2012). As per the results, forskolin enhances the bone formation by activating human mesenchymal stromal cells and also PKA acts as a major string in balancing the adipogenic and osteogenic differentiation. Two PKA activators, i.e., 8-bromo-cAMP and forskolin were compared with standard dibutyryl cyclic adenosine monophosphate (db-cAMP) for their effects on proliferation and osteogenic differentiation. All the three compounds used in the study induced alkaline phosphatase levels, bone-specific target genes, and secretion of insulin-like growth factor-1. Along with this, proliferation of hMSCs was inhibited in a dose-dependent manner, and compared to other two compounds, greater amount of bone formation was initiated by forskolin treatment (Doorn et al. 2012).

5.2.22 Nerve Regeneration

Kilmer and Carlsen (1984) reported that accumulating adenylate cyclase activity was translated into a twofold increase in cAMP concentration in the regenerating nerve stump, coincident with the initiation and elongation of regenerative nerve sprouts.

The role of cAMP in regeneration by using forskolin was studied in situ. In the study, forskolin produced an approximately 40-fold greater elevation in neuronal cAMP than an equimolar (10^{-5}) concentration of isoprenaline. Moreover, the elevated cAMP concentration persisted for at least 60 min in the continued presence of forskolin. Daily injection of forskolin into the dorsal lymph sac of *Rana pipiens*, or delivery of forskolin through an implanted osmotic pump, produced a sustained 40% increase in sensory nerve regeneration rate in freeze-lesioned sciatic nerves. Hence, as per the study, stimulation of nerve growth is achieved by the increase in cAMP concentration, and the activation of appropriate protein kinases.

5.2.23 Functional Neural Differentiation of Human Adipose Tissue-Derived Stem Cells

Adult mesenchymal stem cells (MSCs) derived from adipose tissue are reported to differentiate into mesenchymal, endodermal, and ectodermal cell lineages in vitro. In order to prove the same, the multipotent ability of human adipose tissue-derived stem cells (hADSCs) as MSCs and investigation of their neural differentiation potential was carried out. Treatment with bFGF and forskolin initiated the differentiation of neural cells including neurons and glia. Compared to the primary hADSCs, a significant increase in immunoreactivities, neuronal markers, astrocyte marker, and oligodendrocyte marker was observed in neural differentiated-hADSCs (NI-hADSCs) cells. Moreover, NI-hADSCs also acquired neuron-like functions, and expressed high levels of ionic channel genes for sodium, potassium, and calcium molecules with the treatment. Hence the study, emphasizes the role of bFGF and forskolin in initiating the self-renewing capacity of hADSCs and its multipotency as stem cells (Jang et al. 2010).

5.2.24 Neuronal Acetylcholinesterase Inhibitor

Falé (2011) and Falé et al. (2011, 2012) determined the function of *C. forskohlii* herbal tea as inhibitor of the brain AChE activity. When the plant extract was intragastrically administered, vestigial amounts of metabolites from *C. forskohlii* extract compounds were present in rat plasma, but none were found in brain, although inhibition of brain acetylcholinesterase activity was detected. However, when *C. forskohlii* extract was administered intraperitoneally, all its compounds were found in plasma, and rosmarinic acid was found in brain.

5.2.25 Antipsoriasis Effect

Dean (2012) reported that forskolin successfully cures psoriasis by regulating cAMP levels in skin cells which enables its use in the treatment of psoriasis.

5.2.26 Anticonvulsant Activity

Fernandes et al. (2012) reported the anticonvulsant activity of a hydroalcoholic extract of *C. forskohlii* leaves on seizures induced by strychnine sulfate (2.0 mg/kg) but was quite ineffective against pilocarpine-induced convulsions (600 mg/kg) in mice.

5.2.27 Neurological Disorders

Neuroprotective strategies of blood–brain barrier penetrant "forskolin" (AC/cAMP/PK_A/CREB activator) to ameliorate mitochondrial dysfunctioning in neurotoxic experimental model of autism were investigated by Mehan et al. (2019). They revealed that adenyl cyclase activator, that is, FSK-mediated cAMP/CREB activation, might be a unique platform for the prevention of neurodegenerative diseases.

5.2.28 Immunomodulator Activity

Kapewangolo et al. (2013) investigated the influence of ethyl acetate fraction of *C. forskohlii* leaves on secretion inhibition of proinflammatory cytokines by PBMCs and verified that this extract significantly inhibited the secretion of IL-2, IL-6, IL-10, TNF, and IL-17A at 25 and 50 μg/ml. Cordeiro et al. (2022) evaluated the immunomodulatory potential of *C. forskohlii* (Syn.: *Plectranthus barbatus*) leaf aqueous extract (CFA) and acetone: water (7:3) organic extract (CFO). CFO at 100 $\mu g.mL^{-1}$ significantly inhibited IFN-γ and IL-17A cytokines.

5.2.29 Analgesic Activity

Ezeonwumelu et al. (2019) investigated the analgesic activity of aqueous leaf extract of *C. forskohlii* (Syn.: *Plectranthus barbatus*) at 100, 200, and 400 mg/kg of body weight. Tail-flick test was nonsignificant while formalin-induced pain test demonstrated significant activity.

5.2.30 Antiprotozoal Activity

Chromatographic separation of the *n*-hexane extract of the aerial part of *C. forskohlii* led to the isolation of five abietane-type diterpenes: dehydroabietane (1); 5,6-didehydro-7-hydroxy-taxodone (2); taxodione (3); 20-deoxocarnosol (4); and 6α,11,12,-trihydroxy-7β,20-epoxy-8,11,13-abietatriene (5) (Mothana et al. 2014). The isolated abietane-type diterpenes were tested in vitro for their antiprotozoal activity against erythrocytic schizonts of *Plasmodium falciparum*, intracellular amastigotes of *Leishmania infantum*, *Trypanosoma cruzi*, and free trypomastigotes of *T. brucei*. Compound (2) 5,6-didehydro-7-hydroxy-taxodone showed remarkable activity with acceptable selectivity against *P. falciparum* and *T. brucei*. Compounds (3)–(5) exhibited non-specific antiprotozoal activity due to high cytotoxicity. Dehydroabietane showed no antiprotozoal potential.

Methanolic extract from *C. forskohlii* leaves showed in vitro anti-trypanosomal activity (dos Santos et al. 2012). The bioassay-guided fractionation resulted in the isolation of a gallic acid derivative, identified as 1,2,3,4,6-penta-O-galloyl-{beta}-D-glucose (PGG). This compound was tested against trypomastigote forms of *T. cruzi* and displayed an EC{sub 50} value of 67 {mu}M, at least 6.6-fold more effective than the standard drug benznidazole.

5.2.31 Antimalarial Activity

Lawi et al. (2018) evaluated the antimalarial activity of *C. forskohlii* (Syn.: *Plectranthus barbatus*) extracts. Extracts from leaves are more active against *Anopheles gambiae* with $LC_{50} = 55.65$ μg/mL followed by twigs and roots with $LC_{50} = 465$ μg/mL and $LC_{50} = 636$ μg/mL at 72 h exposure, respectively. Twig extracts showed a moderate effect on both *A. gambiae* and *A. eagyptiae* after 48 h and 72 h exposure time with a 100% (MRC) Mortality Rate Concentration of >1000 μg/mL.

5.2.32 Anthelmintic Activity

Sirama et al. (2019) investigated anthelmintic activity of *C. forskohlii* (Syn.: *Plectranthus barbatus)* (leaves) used by traditional medicine practitioners of Migori County using adult *Haemonchus contortus*. The anthelmintic activities of 6.25 mg/mL, 12.5 mg/mL, and 25 mg/mL concentrations of water, acetone, and methanol crude extracts of *C. forskohlii* (leaves) were compared with the effect produced by the standard drug albendazole. Death of *H. contortus* worm was determined within a period of 24 h. The extract had mean mortality of 0–16.7% at 6.25 mg/mL; 3.3–26.7% at 12.5 mg/mL; and 6.7–33.3% at 25 mg/mL.

5.2.33 Gastroprotective and Anti-*Helicobacter pylori* Effects

Pezzin et al. (2017) reported gastroprotective and anti-*Helicobacter pylori* effects of *C. forskohlii* (Syn.: *Plectrnthus barbatus*). The oral treatment with the extracts showed gastroprotection ranging 60–71%. The anti-*H. pylori* results were significant, showing MIC of 256 μg/mL for the ethyl acetate fraction (EAF) and 512 μg/mL for the aqueous extract. Porfirio et al. (2010) also reported gastroprotective activity of herbal tea of *C. forskohlii.*

5.2.34 Other Uses

Alwyn and White (1992) reported that forskolin acts as a noncompetitive inhibitor of nicotinic acetylcholine receptors. And according to Roger et al. (1987), regulation of dog thyroid epithelial cell cycle was achieved by forskolin and adenylate cyclase activator. Chen et al. (2009) conducted an in vitro skin diffusion study of pure forskolin versus forskolin containing *C. forskohlii* root extract.

5.3 Nano-Emulsions and Nanoparticles as Vehicles for Topical Delivery of Forskolin

Miastkowska et al. (2017) reported the effect of nano-emulsions as vehicles for topical delivery of forskolin topical administration.

Forskolin-loaded human serum albumin nanoparticles (FR-HSANPs) were successfully prepared by incorporation and affinity-binding methods by Nagati et al. (2020). The drug loading was more than 88% and further sustained release profiles were observed as it is 77.5% in 24 h time. The cytotoxicity results with HepG2 cells indicated that FR-HSANPs showed significantly higher cytotoxicity and lower cell viability as compared to free forskolin (FR). Further, Nagati et al. (2020) used circular dichroism and molecular dynamics simulations to elucidate the possible structural changes including local conformational changes and rigidity of the residues of both HSA and HSA-forskolin complexes.

5.4 Intestinal Permeability of Forskolin

The intestinal permeability of forskolin was investigated by Liu et al. (2012) using a Single-Pass Intestinal Perfusion (SPIP) technique in rats. Three different concentrations of forskolin (11.90, 29.75, and 59.90 μg/mL) were used to study the permeability across duodenum, jejunum, ileum, and colon. The results of the study indicated that forskolin could be absorbed in all the segments of the intestine with an effective permeability (P (eff)) comparable to that of drugs with high intestinal permeability. The P (eff) was highest in the duodenum as compared to other intestinal segments and decreased at highest concentration of Forskolin, suggesting a saturation of permeability. Forskolin was found to be a substrate for P-glycoprotein (PgP) as the addition of verapamil (inhibitor of PgP) could enhance its permeability across the rat jejunum. After oral administration in humans, the absorbed fraction of dissolved forskolin was estimated to be 100% calculated from rat P (eff). In conclusion, dissolved forskolin can be absorbed readily in the intestine.

5.5 Effects on Pregnancy

Toxic effects of *C. forskohlii* were investigated during the different periods of pregnancy in rats. Hydroalcoholic extract of *C. forskohlii* at 880 mg/kg/day exerted a variety of toxic effects including delay in fetal development and imposed anti-implantation. After embryo implantation, a delay in the development associated with maternal toxicity was observed (Almeida and Lemonica 2000). Hence, care should be taken in consuming *C. forskohlii* during gestation period.

5.6 Toxicological Studies

Acute, subacute, chronic oral toxicity, and mutagenicity of hydroethanolic extract of *C. forskohlii* (CF), standardized to 10% forskolin was investigated in male and female Wistar rats. A single dose of 2000 mg/kg body weight once daily for 14 consecutive days was evaluated for toxicity, general behavior, and pharmacological effects. In subacute oral toxicity, the test substance was administered for 28 days with daily doses of 100, 300, and 1000 mg/kg body weight. No deaths were reported in all the toxicity studies performed. No significant changes were observed in the hematology and serum biochemistry values from the control group animals. CF extract (10% forskolin) did not produce any significant toxic effects in Wistar rats at 1000 mg/kg body weight and had no potential to induce mutagenicity (Majeed et al. 2015a). Acute toxicity tests by Ezeonwumelu et al. (2019) revealed no deaths in rats after oral treatment with up to 10,000 mg/kg of extract.

A post-marketing nationwide online survey was conducted by Nishijima et al. (2019) for the herbal ingredient *C. forskohlii* extract (CFE). The fitted curve showed that the safe intake amount of CFE was less than 250 mg/day; however, considering its effectiveness, 500 mg/day of CFE might be acceptable.

5.7 Antifeedant Activity

An aphid antifeed diterpene, plectrin isolated from *C. forskohlii* showed antifeedant activity to the green bug *Schizaphis gramineum* and pink bollworm *Pectinophora gossypiella*. A 50% feedant inhibition was observed at 50 and 100 ppm of plectrin (Kubo et al. 1984).

5.8 Patents

A variety of pharmacological activities of *C. forskohlii* were the subjects of a number of patents, for example, lean body mass and energy balance (Majeed 1998, 2005, 2008, 2018a, b, 2019a, b, 2020), antispasmodic effects on smooth muscle of the respiratory system, antiasthmatic, cough-relieving and phlegm-expelling (Zhang et al. 2005; Liu et al. 2006; Jin et al. 1997), inhibition of the absorption of alveolar bones (Tanaka 2003), reduction of the total body weight (Godard et al. 2005; Badmaev and Majeed 2007), induction of lipolysis in rat adipose tissue (Yamashita et al. 2004), and inhibition of the α-glucosidase (Miura 2007) as well as promotion of subcutaneous fat decomposition (Saito et al. 2000). Other patents presented an antiallergic effect (Kawakami 2003), hair-loss preventing effect and activation of the process of melanogenesis (Bonte et al. 1999), an antiaging effect (Adachi et al. 1996) as well as antimicrobial activity of the essential oil (Majeed and Prakash 2003).

References

Abd-Allah HML, AbdAlhady MM, Abdelall MNA, Mohamedy SN, Alharbi WDM, Bahathiq AO (2017) Hormonal and immunological responses to *Coleus forskohlii* treatment in female rats with experimentally polycystic ovaries syndrome. Zagazig Vet J 45(1):74–81

Abe A, Karaki H (1989) Effect of forskolin on cytosolic ca++ level and contraction in vascular smooth muscle. J Pharmacol Exp Ther 249:895–900

Abraham SN, Miao Y (2015) The nature of immune responses to urinary tract infections. Nat Rev Immunol 15(10):655–663

Adachi H, Ehata S, Hayashi T (1996) Antiaging cosmetics containing *Coleus forskohlii* root extracts and antioxidants. Japanese Patent JP 08176005 A

Adnot S, Desmier M, Ferry N, Hanoune J, Sevenet T (1982) Forskolin (a powerful inhibitor of human platelet aggregation). Biochem Pharmacol 31:4071–4074

Agarwal KC, Parks RE (1983) Forskolin: a potential antimetastatic agent. Int J Cancer 32:801–804

Agarwal KC, Zielinski BA, Maitra RS (1989) Significance of plasma adenosine in the antiplatelet activity of forskolin: potentiation by dipyridamole and dilazel. Thrombi Haemost 61:106–110

Alasbahi RH, Melzig MF (2010) *Plectranthus barbatus*: a review of phytochemistry, ethnobotanical uses and pharmacology – part 2. Planta Med 76:753–765

Alasbahi RH, Melzig MF (2012) Forskolin and derivatives as tools for studying the role of cAMP. Parmazie 67:5–13

Al-Ghamdi AY (2021) Phytochemical screening and antimicrobial activity of stem extract of *Coleus forskohlii* L. collected from Al-Baha area, Saudi Arabia. GSC Biol Pharm Sci 16 (2):78–86

Alghamdi MSK, Fakieh M, Anwar Y, Ali HSHM, Al-Matary M, Al-Maaqar SMS, Arif M, Al-Ahmadi TM (2021) Bactericidal activity of *Coleus forskohlii* extract against multi drug resistant *Acinetobacter baumannii* strains isolated from hospital. Biosci Biotechnol Res Commun 14:3

Allen DO, Ahmed B, Naseer K (1986) Relationships between cyclic AMP levels and lipolysis in fat cells after isoproterenol and forskolin stimulation. J Pharmacol Exp Ther 238:659–664

Almeida FC, Lemonica IP (2000) The toxic effects of *Coleus barbatus* on the different periods of pregnancy in rats. J Ethnopharmacol 73:53–60

Alwyn ML, White MM (1992) Forskolin acts as a noncompetitive inhibitor of nicotinic acetylcholine receptors. Mol Pharmacol 41(5):908–913

Ammon H, Müller A (1984) Effect of forskolin on islet cyclic AMP, insulin secretion, blood glucose and intravenous glucose tolerance in rats. Naunyn Schmiedeberg's Arch Pharmacol 326: 364–367

Ammon HP, Müller AB (1985) Forskolin: from an ayurvedic remedy to a modern agent. Planta Med 6:473–477

Anbuselvan S, Muralikrishnan V (2013) Antimicrobial activity of *Coleus forskohlii* root extract against human pathogens. Int J Phytopharmacol 4(1):42–49

Anonymous (2006) Monograph: *Coleus forskohlii*. Altern Med Rev 11(1):47–51. http://www.altmedrev.com/archive/publications/11/1/47.pdf

Atulkar P, Thakur R, Singh P (2015) Antimicrobial activities of root extracts of *Coleus forskohlii* Briq. Intern J Recent Scientific Res 6(12):7732–7734

Badian M, Dabrowski J, Grigoleit HG, Lieb W, Lindner E, Rupp W (1984) Effect of forskolin eyedrops on intraocular pressure in healthy males. Klin Monatsbl Augenheilkd 185:522–526. [in German]

Badmaev V, Majeed H (2004) *Coleus forskohlii*: fat fighting and more. Nutrition Industry Executive pp.1–6 https://drmajeed.com/pdf/articles/2004ColeusForskohlii.pdf

Badmaev V, Majeed H (2007) Compositions for physiological increase of male and female hormones with diterpene forskolin and its derivatives. Patent Application No. AU 2007200442 A1

Badmaev V, Majeed M, Conte A, Parker JE (2002) Diterpene forskolin (*Coleus forskohlii* Benth.): a possible new compound for reduction of body weight by increasing lean body mass. Nutraceuticals, March–April 2002, pp 6–7

Baskaran C, Rathabai V, Sivamani P, Thiagarajan V (2011) Antimicrobial activity of various root extracts of *Coleus forskohlii*. Int J Curr Sci 1:78–84

Battochio AP, Sartori MS, Coelho CA (2005) Water-soluble extract of *Coleus barbatus* modulates weight gain, energy utilization and lipid metabolism in secondary biliary cirrhosis: an experimental study in young rats. Acta Cir Bras 20:229–236

Bauer K, Dietersdorder F, Sertl K, Kaik B, Kaik G (1993) Pharmacodynamic effects of inhaled dry powder formulations of fenoterol and colforsin in asthma. Clin Pharmacol Ther 53:76–83

Baumann G, Felix S, Sattelberger U, Klein G (1990) Cardiovascular effects of forskolin (HL 362) in patients with idiopathic congestive cardiomyopathy—a comparative study with dobutamine and sodium nitroprusside. J Cardiovasc Pharmacol 16:93–100

Becker DE (2013) Antithrombotic drugs: Pharmacology and implications for dental practice. Anesth Prog 60:72–80

Belfrage P, Fredrikson G, Olsson H, Strilfors P (1982) Hormonal regulation of adipose tissue lipolysis by reversible phosphorylation of hormone-sensitive lipase. Prog Clin Biol Res 102: 213–223

Bersudsky Y, Kotler M, Shifrin M, Belmaker RH (1996) A preliminary study of possible psychoactive effects of intravenous forskolin in depressed and schizophrenic patients. J Neural Transm 103:1463–1467

Bhat SV, Dohadwalla AN, Bajwa BS, Dadkar NK, Dornauer H, de Souza NJ (1983) The antihypertensive and positive inotropic diterpene forskolin: effects of structural modifications on its activity. J Med Chem 26:486–492

Bhowal M, Mehta DM (2017) *Coleus forskohlii*: phytochemical and pharmacological profile. Int J Pharm Sci Res 1:3599–3618

Bodiwala HS, Sabde S, Mitra D, Bhutania KK, Singha IP (2009) Anti-HIV diterpenes from *Coleus forskohlii*. Nat Prod Commun 4(9):1173–1175

Boj SF, Vonk AM, Statia M, Su J, Dekkers JF, Vries RR, Beekman JM, Clevers H (2017) Forskolin-induced swelling in intestinal organoids: an in vitro assay for assessing drug response in cystic fibrosis patients. J Vis Exp 120:e55159

Bonte F, Meybeck A, Marechal C (1999) Composition based on hydrated lipidic lamellar phases or on liposomes containing at least one derivative of labdane, or a plant extract containing it; cosmetic or pharmaceutical, particularly dermatological composition containing it. US Patent 5891464

Borges AS, Minozzo BR, Santos H, Ardisson JS, Rodrigues RP, Romao W, Borges WS, Gonçalves RCR, Beltrame FR, Kitagawa RR (2020) *Plectranthus barbatus* Andrews as anti-*Helicobacter pylori* agent with activity against adenocarcinoma gastric cells. Ind Crop Prod 146:1–12. https://doi.org/10.1016/j.indcrop.2020.112207

Bristow M, Strosberg A, Ginsburg R (1983) Forskolin activation of human myocardial adenylatecyclase, circulation. Amer Heart Assoc, Dallas, TX, USA

Bristow MR, Ginsburg R, Strosberg A, Montgomery W, Minobe W (1984) Pharmacology and inotropic potential of forskolin in the human heart. J Clin Invest 74:212–223

Brubaker RF, Carlson KH, Kullerstrand LJ, McLaren JW (1987) Topical forskolin (colforsin) and aqueous flow in humans. Arch Ophthalmol 105:637–641

Bruka JF (1986) Forskolin: its chemical biological and medical potential. In: Proc int Symp forskolin. Hoechst India Ltd., Bombay, pp 117–136

Burns TW, Langley PE, Terry BE, Bylund DB, Forte LR Jr (1987) Comparative effects of forskolin and isoproterenol on the cyclic AMP content of human adipocytes. Life Sci 40:145–154

Câmara CC, Nascimento NR, Macedo-Filho CL, Almeida FB, Fonteles MC (2003) Antispasmodic effect of the essential oil of *Plectranthus barbatus* and some major constituents on the Guinea-pig ileum. Planta Med 69:1080–1085

Caprioli J, Sears M (1983) Forskolin lowers intraocular pressure in rabbits, monkeys and man. Lancet 30:958–960

Caprioli J, Sears M, Bausher L, Gregory D, Mead A (1984) Forskolin lowers intraocular pressure by reducing aqueous inflow. Invest Ophthalmol Vis Sci 25:268–277

Chakraborty A, Haque SM, Dey D, Mukherjee S, Ghosh B (2022) Detection of UTI pathogen-killing properties of *Coleus forskohlii* from tissue cultured in vitro and ex vitro plants. Proc Natl Acad Sci India Sect B Biol Sci 92:157–169. https://doi.org/10.1007/s40011-021-01285-4

Chen J, Hammell DC, Spry M, D'Orazio JA, Stinchcomb AL (2009) In vitro skin diffusion study of pure forskolin versus a forskolin-containing *Plectranthus barbatus* root extract. J Nat Prod 72(4):769–771

Chen J-Y, Peng S-Y, Cheng Y-H, Lee I-T, Yu YH (2021) Effect of forskolin on body weight, glucose metabolism and adipocyte size of diet-induced obesity in mice. Animals 11:645. https://doi.org/10.3390/ani11030645

Christenson JT, Thulesius O, Al-Huneidi W (1989) Forskolin inhibits platelet deposition in small diameter polytetrafluoroethylene grafts. J Cardiovasc Surg 30(1):81–84

Christenson JT, Thulesius O, Nazzal MM (1995) The effect of forskolin on blood flow, platelet metabolism, aggregation and ATP release. Vasa 24(1):56–61

Christenson JT, Thulesius O, Owunwanne A, Nazzal MM (1991) Forskolin impregnation of small calibre PTFE grafts lowers early platelet graft sequestration and improves patency in a sheep model. Eur J Vas Surg 5(3):271–275

Cordeiro MF, Nunes TRS, Bezerra FG, Damasco PKM, Silva WAV, Ferreira MRA, Magalhães OMC, Soares LAL, Cavalcanti IMF, Pitta MGR, Rêgo MJBM (2022) Phytochemical characterization and biological activities of *Plectranthus barbatus* Andrews. Brazilian J Biol 82. https://doi.org/10.1590/1519-6984.236297

Costa MCCD, Nascimento SC (2003) Actividade citotóxica de *Plectranthus barbatus* Andr. (Lamiaceae). Acta Farm Bonaer 22:155–158

Damera S, Ajmera RR, Ciddi V (2019) Forskolin alleviates diabetic nephropathy via inhibition of aldose reductase and advanced glycation end products formation. Diabet Updates 5:1–5. https://doi.org/10.15761/DU.1000127

De Boeck K, Amaral MD (2016) Progress in therapies for cystic fibrosis. Lancet Respir 4:662–674

de Chaffoy de Courcelles D, Roevens P, Van Belle H (1987) Prostaglandin E1 and forskolin antagonize C-kinase activation in the human platelet. Biochem J 244:93–99

De Freitas AA, Lima NM, Santos IV, de Castro SBR, Pereira PS, Santos LL, Rodrigues JL, de Souza Alves CC, Carli AP (2018) Bioactive proteins from *Plectranthus barbatus* and detection by RP-HPLC-PDA. J Pharmacogn Phytochem 7(2):305–309

De Souza NJ (1993) Industrial development of traditional drugs: the forskolin example. A mini-review. J Ethnopharmacol 38:177–180

De Souza NJ, Dohadwalla AN, Reden J (1983) Forskolin: a labdane diterpenoid with antihypertensive, positive inotropic, platelet aggregation inhibitory and adenylate cyclase activating properties. Med Res Rev 3:201–219

Dean W (2012) Forskolin and cAMP. NaturoDoc.com http://www.naturodoc.com/forskolin.htm

Ding X, Staudinger JL (2005) Induction of drug metabolism by forskolin: the role of the pregnane X receptor and the protein kinase a signal transduction pathway. J Pharmacol Exp Ther 312(2): 849–856

Dohadwalla AN (1985) Biological activities of forskolin. In: Rupp RH, de Souza NJ, Dohadwalla AN (eds) Forskolin: its chemical biological and medical potential. Hoechst India, 1986, Bombay, pp 19–30

Doni MG, Deana R, Bertoncello S, Zoccarato F, Alexandre A (1988) Forskolin and prostacyclin inhibit fluoride induced platelet activation and protein kinase C dependent responses. Biochem Biophys Res Commun 156:1316–1323

Doorn J, Siddappa R, Van Blitterswijk CA, De Boer J (2012) Forskolin enhances in vivo bone formation by human mesenchymal stromal cells. Tissue Eng Part A 18:558–567

dos Santos RT, Hiramoto LL, Lago JH, Sartorelli P, Tempone AG, Pinto EG (2012) Anti-trypanosomal activity of 1,2,3,4,6-penta-O-galloyl-β-D-glucose isolated from *Plectranthus barbatus* Andrews (Lamiaceae). Quim Nova 35:11. https://doi.org/10.1590/S0100-40422012001100025

Doseyici S, Mehmetoglu I, Toker A, Yerlikaya FH, Erbay E (2014) The effects of forskolin and rolipramon cAMP, cGMP and free fatty acid levels in diet induced obesity. Biotech Histochem 89:388–392

Dowless MS, Barbee JL, Borchert KM, Bocchinfuso WP, Houck KA (2005) Cyclic AMP-independent activation of CYP3A4 gene expression by forskolin. Eur J Pharmacol 512(1):9–13

Drumm ML, Wilkinson DJ, Smit LS, Worrell RT, Strong TV, Frizzell RA, Dawson DC, Collins FS (1991) Chloride conductance expressed by delta f508 and other mutant cftrs in xenopus oocytes. Science 254:1797–1799

Dubey MP, Srimal RC, Patnaik GK Dhawan BN (1974) Blood pressure lowering and antispasmodic effects of *C. forskohlii* Briq. Indian J Pharm 6:15

Dubey MP, Srimal RC, Nityanand S, Dhawan BN (1981) Pharmacological studies on colenol, a hypotensive diterpene from *Coleus forskohlii*. J Ethnopharmacol 3:1–13

El-Agroudy NN, El-Naga RN, El-Razeq RA, El-Demerdash E (2016) Forskolin, a hedgehog signalling inhibitor, attenuates carbon tetrachloride-induced liver fibrosis in rats. Br J Pharmacol 173:3248–3260

Ezeonwumelu J, Kawooya G, Okoruwa A, Dare S, Ebosie J, Akunne A, Tanayen J, Udechukwu B (2019) Phytochemical screening, toxicity, analgesic and anti-pyretic studies of aqueous leaf extract of *Plectranthus barbatus* [Andrews. Engl.] in rats. Pharmacol Pharm 10:205–221

Fadlelmula AA, Al-Ghamdi AY, Abdalla MOM (2020) In vitro antioxidant activity, total phenolic content and antimicrobial activity of *Coleus forskohlii* in Al-Baha, Saudi Arabia. Asian J Chem 32(8):2033–2037

Falé PLV (2011) Biological activities of *Plectranthus barbatus* aqueous extracts. in vitro and in vivo studies of activity, bioavailability and metabolism. PhD theis, Universidade de Lisboa

Falé PLV, Ascensão L, Serralheiro MLM, Haris P (2012) Interaction between *Plectranthus barbatus* herbal tea compounds and acetylcholinesterase: binding and activity studies. Food Funct 3:1176–1184

Falé PLV, Madeira PJA, Florêncio MH, Luísa M, Serralheiro M (2011) Function of *Plectranthus barbatus* herbal tea as neuronal acetylcholinesterase inhibitor. Food Nutrition 2:130–136

Feng S, Wang L, Voravuthikunchai S, Liang Z, Liu A, Bao F (2016) Isoforskolin from native plant *Coleus forskohlii* of Yunnan, China plays multiple biological roles. Open J Immunol 6:63–66. https://doi.org/10.4236/oji.2016.62006

Fernandes LCB, Camara CC, Soto-Blanco B (2012) Anticonvulsant activity of extracts of *Plectranthus barbatus* leaves in mice. Evid Based Complement Alternat Med 2012:860153. https://doi.org/10.1155/2012/860153

Figueiredo NL, Aguiar SRMM, Falé PL, Ascensão L, Serralheiro MLM, Lino ARL (2010) The inhibitory effect of *Plectranthus barbatus* and *Plectranthus ecklonii* leaves on the viability, glucosyl transferase activity and biofilm formation of *Streptococcus sobrinus* and *Streptococcus mutans*. Food Chem 119:664–668

Figueiredo NL, Fale PL, Madeira PJA, Florencio MH, Ascensao L, Serralheiro MLM, Lino ARL (2014) Phytochemical analysis of *Plectranthus* sp. extracts and application in inhibition of dental bacteria, *Streptococcus sobrinus* and *Streptococcus mutans*. Eur J Med Plants 4(7): 794–809

Fischman LA, Skorupa LA, Souccar C, Lapa AJ (1991) The water extract of *Coleus barbatus* Benth decreases gastric secretion in rats. Mem Inst Oswaldo Cruz 86(Suppl. 2):141–143

Ganash M, Qanash S (2018) Phenolic acids and biological activities of *Coleus forskohlii* and *Plectranthus barbatus* as traditional medicinal plants. Int J Pharmacol 14:856–865

Gelmini F, Squillace P, Testa C, Sparacino AC, Angioletti S, Beretta G (2015) GC–MS characterisation and biological activity of essential oils from different vegetative organs of *Plectranthus barbatus* and *Plectranthus caninus* cultivated in North Italy. Nat Prod Res 29(11): 993–998

Geng S, Sun B, Liu S, Wang J (2007) Up-regulation of connexin 43 and gap junctional intercellular communication by Coleusin factor is associated with growth inhibition in rat osteosarcoma UMR106 cells. Cell Biol Int 31:1420–1427

Geng S, Sun B, Lu R, Wang J (2014) Coleusin factor, a novel anticancer diterpenoid, inhibits osteosarcoma growth by inducing bone morphogenetic protein-2–dependent differentiation. Mol Cancer Ther 13(6):1431–1441

Girardier L (1983) Brown fat: an energy dissipating tissue. In: Girardier L, Sock MJ (eds) Mammalian thermogenesis. Chapman and Hall Ltd., London, New York, pp 50–98

Godard MP, Johnson BA, Richmond SR (2005) Body composition and hormonal adaptations associated with forskolin consumption in overweight and obese men. Obes Res 13:1335–1343

González-Sánchez R, Trujillo X, Trujill-Hernandez B, Vasqiez C, Huerta M, Elizalde A (2006) Forskolin versus sodium cromoglycate for prevention of asthma attacks: a single-blinded clinical trial. J Int Med Res 34:200–207

Graber SE, Hawiger J (1982) Evidence that changes in platelet cyclic AMP levels regulate the fibrinogen receptor on human platelets. J Biol Chem 257:14606–14609

Gupta S, Samanta MK (2010) Design and evaluation of thermoreversible in situ gelling system of forskolin for the treatment of glaucoma. J Pharm Dev Technol 15:386–393

Gupta S, Samanta MK, Raichur AM (2010) Dual-drug delivery system based on in situ gel-forming nanosuspension of forskolin to enhance antiglaucoma efficacy. AAPS PharmSciTech 11:322–335

Han LK, Morimoto C, Yu RH, Okuda H (2005) Effects of *Coleus forskohlii* on fat storage in ovariectomized rats. Yakugaku Zasshi 125:449–453

Hayashi T, Hikichi M, Yukitake J, Wakatsuki T, Nishio E, Utsumi T, Harada N (2018) Forskolin increases the effect of everolimus on aromatase inhibitor-resistant breast cancer cells. Oncotarget 9:23451–23461

Hayley L (2015) *C. forskholii* extract supplementation in conjunction with a hypocaloric diet reduces the risk factors of metabolic syndrome in overweight and obese subjects: a randomized controlled trial. Nutrients 7:9508–9522

Henderson S, Magu B, Rasmussen C, Lancaster S, Kerksick C, Smith P, Melton C, Cowan P, Greenwood M, Earnest C, Almada A, Milnor P, Magrans T, Bowden R, Ounpraseuth S, Thomas A, Kreider RB (2005) Effects of *Coleus forskohlii* supplementation on body composition and hematological profiles in mildly overweight women. J Int Soc Sports Nutr 9:54–62

Hiramatsu T, Kume H, Kotlikoff MI, Takagi K (1994) Role of calcium-activated potassium channels in the relaxation of tracheal smooth muscles by forskolin. Clin Exp Pharmacol Physiol 21:367–375

Ho R-J, Shi Q-H (1982) Forskolin as a novel lipolytic agent. Biochem Biophys Res Commun 107: 157–164. https://doi.org/10.1016/0006-291X(82)91683-7

Holz GG (2004) Epac: a new camp-binding protein in support of glucagon-like peptide-1 receptor-mediated signal transduction in the pancreatic cell. Diabetes 53:5–13

Huang DH, Yang YF, Ali LQ, Lu Y, Wu HZ (2011) Studies on the chemical constituents of *Coleus forskohlii* transplanted in Tongcheng and their antitumor activity. Zhong Yao Cai 34(3):375–378

Ibrahim MM, Arafa NM, Aly UI (2018) Antioxidant activity, phenol and flavonoid contents of plant and callus cultures of *Plectranthus barbatus* Andrews. Egypt Pharmaceut J 17:32–39

Illiano M, Sapio L, Caiafa I, Chiosi E, Spina A, Naviglio S (2017) Forskolin sensitizes pancreatic cancer cells to gemcitabine via Stat3 and Erk1/2 inhibition. AIMS Mol Sci 4(2):224–240

Illiano M, Sapio L, Salzillo A, Capasso L, Caiafa I, Chiosi E, Spina A, Naviglio S (2018) Forskolin improves sensitivity to doxorubicin of triple negative breast cancer cells via protein kinase A-mediated ERK1/2 inhibition. Biochem Pharmacol 152:104–113

Iwase M, Ishikawa Y, Shen YT, Shannon RP, Sato N, Ganguly PK, Eki T, Vatner DF, Vatner SF (1996) Neurally mediated cardiac effects of forskolin in conscious dogs. Am J Phys 271: H1473–H1482

Jagtap M, Chandola HM, Ravishankar B (2011) Clinical efficacy of *C.forskholii* (Wild) Briq (Makandi) in hypertension of geriatric population. AYU 32(1):59–65

James P (2003) *Coleus forskohlii*: a nonstimulant herb with proven fat-burning ability. Dynamic Chiropractic 21:12

Jang S, Cho H-H, Cho Y-B, Park J-S, Jeong H-S (2010) Functional neural differentiation of human adipose tissue-derived stem cells using bFGF and forskolin. BNC Cell Biol 11:25. https://doi.org/10.1186/1471-2121-11-25

Jeff LS, Xunshan D, Kristin L (2006) Pregnane X receptor and natural products: beyond drug-drug interactions. Expert Opin Drug Metab Toxicol 2:847–857

Jin Q, Chen Z, Zhang G (1997) A Chinese medicinal granule for treating bronchial asthma containing flos colei esquirolii, radix peucedani and radix glycyrrhizae. Chinese Patent CN 1154861 A

Joost HG, Habberfield AD, Simpson IA, Laurenza A, Seamon KB (1988) Activation of adenylate cyclase and inhibition of glucose transport in rat adipocytes by forskolin analogues: structural determinants for distinct sites of action. Mol Pharmacol 33:449–453

Joost HG, Steinfelder HJ (1987) Forskolin inhibits insulin-stimulated glucose transport in rat adipose cells by a direct interaction with the glucose transporter. Mol Pharmacol 31(3):279–283

Kaik G, Witte PU (1986) Protective effect of forskolin against acetylcholine provocation in Wochenschr 136: 637–641

Kala S (2014) Antimicrobial activity of *Coleus forskohlii* (Willd.) Briq. and *Costus igneus* n.e.Br. IOSR J Pharm Biol Sci 9(5):1–6

Kamini K, Ashashri S, Gahunge P, Nagar L (2013) A comprehensive review: *Coleus forskohlii*. Int J Ayurved Herbal Med 3(1):1106–1113

Kamohara S (2016) An evidence based review: anti-obesity effects of *Coleus forskohlii*. Pers Med Universe 5:16–20

Kapewangolo P, Hussein AA, Meyer D (2013) Inhibition of HIV-1 enzymes, antioxidant and anti-inflammatory activities of *Plectranthus barbatus*. J Ethnopharmacol 149:184–190. https://doi.org/10.1016/j.jep.2013.06.019

Kariya T, Morito F, Sakai T, Takahata K, Yamanaka M (1985) Effect of forskolin on platelet deaggregation and cyclic AMP generation. Naunyn Schmiedeberg's Arch Pharmacol 331:119–121

Kasonia K (1995) Preliminary screening of plant extracts used in respiratory pathology in Kivu/Zaire on isolated Guinea pigs rings trachea. Belgian J Bot 128:165–175

Kaushal K, Shinde A, Gahunge PNL (2013) A comprehensive review: *Coleus forskohlii*. Int J Ayurvedic Herbal Med 3(1):1106–1113

Kavitha C, Rajamani K, Vadivel E (2010) *Coleus forskohlii*: a comprehensive review on morphology, phytochemistry and pharmacological aspects. J Med Plants Res 4(4):278–285

Kawakami T (2003) Anti-allergy compositions containing *Coleus forskohlii* root extract and cinnamon extract. Japanese Patent JP 2003252786 A

Khan N, Ameeduzzafar KK, Bhatnagar A, Ahmad FJ, Ali A (2018) Chitosan-coated PLGA nanoparticles amplify the ocular hypotensive effect of forskolin: statistical design, characterization and in vivo studies. Int J Biol Macromol 116:648–663

Khatun S (2020) Antimicrobial activity of tuber extracts of the medicinal plant *Coleus forskohlii*. Plant cell Biotechnol Mol Biol 21(11–12):11–17

Khatun S, Chatterjee NC, Cakilcioglu U (2011) Antioxidant activity of the medicinal plant *Coleus forskohlii* Briq. Afr J Biotechnol 10(13):2530–2535

Kilmer SL, Carlsen RC (1984) Forskolin activation of adenylate cyclase in vivo stimulates nerve regeneration. Nature 307:455–457

Kim HK, Song KS, Chung J-H, Lee KR, Lee S-N (2004) Platelet microparticles induce angiogenesis in vitro. Br J Haematol 124:376–384

Kramer W, Thormann J, Kindler M, Schlepper M (1987) Effects of forskolin on left ventricular function in dilated cardiomyopathy. Arzneimittelforschung 37:364–367

Kreuter W, Chapman RW, Gulbenkian A, Tozzi S (1985) Bronchodilator and antiallergy activity of forskolin. Eur J Pharmacol 111:1–8

Krieder R, Henderson S, Magu B et al (2005) Effects of *Coleus forskohlii* supplementation on body composition and haematological profiles in mildly overweight women. J Int Soc Sports Nutr 2 (2):54–62

Kubo I, Matsumoto T, Tori M, Asakawa Y (1984) Structure of plectrin, an aphid antifeedant diterpene from *Plectranthus barbatus*. Chem Lett 9:1513–1516

Lakshmanan GMA, Manikandan S (2015) Review on pharmacological effects of *Plectranthus forskohlii* (Willd) Briq. Int Letters Nat Sci 1:1–9

Lakshmanan GMA, Manikandan S, Panneerselvam S (2013) *Plectranthus forskohlii* (wild) Briq. (Syn: *Coleus forskohlii*)—a compendium on its botany and medicinal uses. Int J Res Plant Sci 3:72–80

Laurberg P (1984) Forskolin stimulation of thyroid secretion of t4 and t3. FEBS Lett 170:273–276

Laurenza A, Sutkowski EM, Seamon KB (1989) Forskolin: a specific stimulator of adenyl cyclase or a diterpene with multiple sites of action? Trends Pharmacol Sci 10(11):442–447

Lawi Y, Sari J, Kidukuli AW (2018) Brine shrimp cytotoxicity, phytochemical screening and larvicidal activities of *Plectranthus barbatus* extracts. Res Rev Insights 2(1):1–4. https://doi.org/10.15761/RRI.1000130

Lee PY, Podos SM, Serle JB, Camras CB, Severin CH (1987) Intraocular pressure effects of multiple doses of drugs applied to glaucomatous monkey eyes. Arch Ophthalmol 105:249–252

Li S, Yang QR, Wang XM, Zou GA, Liu YW (2006) Chemical constituents of *Coleus forskohlii* replanted to Tongcheng. Zhongcaoyao 37:824–826

Lichey J, Friedrich T, Priesnitz M, Biamino G, Usinger P, Huckauf H (1984) Effect of forskolin on methacholine induced broncho constriction in extrinsic asthmatics. Lancet 21:167

Lindner E, Dohadwalla AN, Bhattacharya BK (1978) Positive inotropic and blood pressure lowering activity of a diterpene derivative isolated from *Coleus forskohlii*: forskolin. Drug Res 28:284–289

Lindner E, Metzger H (1983) The action of forskolin on muscle cells is modified by hormones, calcium ions and calcium antagonists. Arzneimittelforschung 33:1436–1441

Litosch I, Hudson TH, Mills I, Li SY, Fain JN (1982) Forskolin as an activator of cyclic AMP accumulation and lipolysis in rat adipocytes. Mol Pharmacol 22(1):109–115

Liu J, Zhang J, Chen Z, Xu YL, Jin QD (2006) Chinese medicine composition for treating asthma and cough and its preparation. Chinese Patent CN 1872146 A

Liu Y, Wu H, Wang XM, Liu J, Xing X (2007) Use of coleon C extracted from Coleus as inhibitor for tumor growth and tumor cell proliferation. Chinese Patent CN 1899273 A

Liu Z-J, Jiang D-B, Tian L-L, Yin J-J, Huang J-M, Weng W-Y (2012) Intestinal permeability of forskolin by in situ single pass perfusion in rats. Planta Med 78:698–702

Loftus H, Astell K, Mathai M, Su X (2015) *Coleus forskohlii* extract supplementation in conjunction with a hypocaloric diet reduces the risk factors of metabolic syndrome in overweight and obese subjects: a randomized controlled trial. Nutrients 7:9508–9522

Lokesh B, Deepa R, Divya K (2018) Medicinal coleus (*Coleus forskohlii* Briq.): a phytochemical crop of commercial significance. J Pharmacogn Phytochem 7:2856–2864

Lygia AF, Skorupa LA, Caden S, Lapa AJ (1991) The water extract of *Coleus barbatus* Benth. decreases gastric secretion in rats. Em Inst Oswaldo Cruz Rio de Janeiro 86:1411–1143

Maeda H, Ozawa H, Saito T, Irie T, Takahta N (1997) Potential antidepressant properties of forskolin and a novel water-soluble forskolin (NKH477) in the forced swimming test. Life Sci 61:2345–2342

Maioli AM, Larissa CA, Andre LC, Michele CL, Daniel JD, Fabio EM (2010) Iron chelating-mediated antioxidant activity of *Plectranthus barbatus* extract on mitochondria. Food Chem 122(1):203–208

Majeed M (1998) Method of preparing a forskohlin composition from forskohlin extract and use of forskohlin for promoting lean body mass and treating mood disorders. US patent 5804596

Majeed M (2005) Forskolin for promoting Lean body mass. European patent No. EP0977564

Majeed M (2008) Forskolin for promoting Lean body mass. Canadian patent No. CA2281562

Majeed M (2012) *Coleus forskohlii* extract in the management of obesity. In: Bagchi D, Preuss HG (eds) Obesity: epidemiology, pathophysiology and prevention, 2nd edn. Taylor & Francis, pp 703–728

Majeed M (2018a) Process and compositions for achieving mammalian energy balance. US patent No. US10085963

Majeed M (2018b) Process and compositions for achieving mammalian energy balance. Australian patent No. 2015351423

Majeed M (2019a) Process and compositions for achieving mammalian energy balance. New Zealand patent NZ720422

Majeed M (2019b) Process and compositions for achieving mammalian energy balance. Japanese patent No. 6472453

Majeed M (2020) Process and compositions for achieving mammalian energy balance. European patent No. EP3194575

Majeed M, Badmaev V, Khan N, Prakash L, Kalyanam N (2009) A new class of phytonutrients for body weight management. Nutra foods 8(1):17–26

Majeed M, Majeed S, Nagabhushanam K, Gnanamani M, Mundkur L (2021) Lesser investigated natural ingredients for the management of obesity. Nutrients 13(2):510. https://doi.org/10.3390/nu13020510

Majeed M, Nagabhushanam K, Natarajan S, Vaidyanathan P, Karri SK (2014a) A double-blind, randomized clinical trial to evaluate the efficacy and safety of forskolin eye drops 1% in the treatment of open angle glaucoma—a comparative study. J Clin Trials 4:1000184

Majeed M, Nagabhushanam K, Natarajan S, Vaidyanathan P, Karri SK (2014b) Intraocular pressure lowering activity of Ocufors (R) (forskolin 0.15% w/v ophthalmic solution) in water loaded New Zealand white rabbits. Int J Pharma Biosci 5(4):328–336

Majeed M, Nagabhushanam K, Natarajan S, Bani S, Vaidyanathan P, Majeed S, Karri SK (2015a) Investigation of acute, sub-acute, chronic oral toxicity and mutagenicity of *Coleus forskohlii* Briq. Hydroethanolic extract, standardized for 10% forskolin in experimental animals. Int J Pharm Pharm Res 5:219–238

Majeed M, Nagabhushanam K, Natarajan S, Vaidyanathan P, Karri SK, Jose JA (2015b) Efficacy and safety of 1% forskolin eye drops in open angle glaucoma—an open label study. Saudi J Ophthalmol 29:197–200

Majeed M, Prakash L (2007) Fighting acne and more effective natural approaches to skin care. Cosmet Toilet 215–219

Majeed M, Prakash S (2003) Composition and methods containing an antimicrobial essential oil extended from *Coleus forskohlii*. US Patent 6607712 B2

Malleswari D, Bagyanarayana G, Hindumathi A (2013) Anti-bacterial activity of *Coleus forskohlii* extracts against some pathogenic bacteria. J Nat Prod Plant Resour 3(4):75–78

Manikandan D, Jeyachandran V, Manikandan A (2018) Synthesis and characterization of silver nanoparticles from *Plectranthus barbatus* seed extract. Int J Pure Appl Mathematics 119(12): 4067–4076

Mathur A, Bhat R, Prasad GBKS, Dua VK (2011) *Coleus barbatus* as a potent antimicrobial agent against some gastro-intestinal pathogens. J Life Sci 3(2):137–140

Matsumoto S, Yamashita T, Araie M, Kametani S, Hosokawa T, Takase M (1990) The ocular penetration of topical forskolin and its effects on intraocular pressure, aqueous flow rate and cyclic AMP level in the rabbit eye. Jpn J Ophthalmol 34:428–435

Matthews RP, McKnight GS (1996) Characterization of the camp response element of the cystic fibrosis transmembrane conductance regulator gene promoter. J Biol Chem 271:31869–31877

Matu EN, van Staden J (2003) Antibacterial and anti-inflammatory activities of some plants used for medicinal purposes in Kenya. J Ethnopharmacol 87:35–41

Mehan S, Khera H, Sharma R (2019) Neuroprotective strategies of blood-brain barrier penetrant "forskolin" (AC/cAMP/PK_A/CREB activator) to ameliorate mitochondrial dysfunctioning in neurotoxic experimental model of autism. In: Borreca A (ed.) Recent Advances in Neurodegeneration. InTech Open. DOI: https://doi.org/10.5772/intechopen.80046

Menon DB, Latha K (2011) Phytochemical screening and *in vitro* anti-inflammatory activity of the stem of *Coleus forskohlii*. Pharmacog J 3(23):75–79

Metzer H, Lindner E (1981) The positive inotropic-acting forskolin, a potent adenylate cyclase activator. Arzneimittelforschung 31(8):1248–1250

Meyer BH, Stulting AA, Müller FO, Luus HG, Badian M (1987) The effects of forskolin eye drops on intra-ocular pressure. S Afr Med J 71:570–571

Miastkowska M, Sikora E, Lasoń E, Garcia-Celma MJ, Escribano-Ferrer E, Solans C, Llinas M (2017) Nano-emulsions as vehicles for topical delivery of forskolin. Acta Biochim Pol 64:713–718

Michael PG, Brad AJ, Scott RR (2005) Body composition and hormonal adaptations associated with forskolin consumption in overweight and obese men. Obesity Res 13:1335–1343

Mitra M, Gantait S, Mandal N (2020) *Coleus forskohlii*: advancements and prospects of in vitro biotechnology. Appl Microbiol Biotechnol 104(6):2359–2371. https://doi.org/10.1007/s00253-020-10377-6

Miura M (2007) Foods containing α-glucosidase inhibitors for restricting calories. Japanese Patent JP 2007195510 A

Mothana RA, Al-Said MS, Al-Musayeib NM, El Gamal AA, Al-Massarani SM, Al-Rehaily AJ, Abdulkader M, Maes L (2014) In vitro antiprotozoal activity of abietane diterpenoids isolated from *Plectranthus barbatus* Andr. Int J Mol Sci 15(5):8360–8371

Mothana RA, Khaled JM, El-Gamal AA, Noman OM, Kumar A, Alajmi MF, Al-Rehaily AJ, Al-Said MS (2019) Comparative evaluation of cytotoxic, antimicrobial and antioxidant activities of the crude extracts of three *Plectranthus* species grown in Saudi Arabia. Saudi Pharmaceut J 27:162–170

Mwitari PG, Ayeka PA, Ondicho J et al (2013) Antimicrobial activity and probable mechanisms of action of medicinal plants of Kenya: *Withania somnifera, Walburgia ugandensis, Prunus africana* and *Plectrunthus barbatus*. PLoS One 8(6):e65619

Nagarajappa SH, Pandit S, Divanji M, Marianna B, Kumar P, Godavarthi A (2016) Effect of *Coleus forskohlii* and its major constituents on cytochrome P450 induction. J Tradit Complement Med 6:130–133

Nagati V, Nakkka S, Yeggoni DP, Subramanyam R (2020) Forskolin-loaded human serumalbumin nanoparticles and its biological importance. J Biomol Struct Dyn 5:1–12

Nguta JM, Kiraithe MN (2019) In vitro antimicrobial activity of aqueous extracts of *Ocimum suave* Willd., *Plectranthus barbatus* Andrews and *Zanthoxylum chalybeum* Engl. against selected pathogenic bacteria. Biomed Biotechnol Res J 3(1):30–34

Nidiry ESJ, Ganeshan G, Lokesha AN (2015) Antifungal activity of the extractives of *Coleus forskohlii* roots and forskolin. Pharm Chem J 49:624–626

Nilani P, Duraisamy B, Dhanabal PS, Khan S, Suresh B, Shankar V, Kavitha KY, Syamala G (2006) Antifungal activity of some *Coleus* species growing in Nilgiris. Ancient Sci Life 26:81–84

Nisar S, Hanif MA, Soomro K, Jilani MI, Kala CP (2020) Chapter 11. Coleus. In: Hanif MA, Nawaz H, Khan MM, Byrne HJ (eds) Medicinal plants of South Asia: novel sources for drug discovery. Elsevier, pp 136–147

Nishijima C, Chiba T, Sato Y, Umegaki K (2019) Nationwide online survey enables the reevaluation of the safety of *Coleus forskohlii* extract intake based on the adverse event frequencies. Nutrients 11(4):866. https://doi.org/10.3390/nu11040866

Okuda H, Morimoto C, Tsujita T (1992) Relationship between cyclic AMP production and lipolysis induced by forskolin in rat fat cells. J Lipid Res 33:225–231

Patel MB (2010) Forskolin: a successful therapeutic pytomolecule. East and Central African J Pharm Sci 13:25–32

Patel T, Saraf M (2016) A review on the pharmacology of *Coleus forskohlii* Briq: a threatened medicinal plant. https://www.researchgate.net/publication/310463016_A_review_on_the_pharmacology_of_coleus_forskohlii_briqa_threatened_medicinal_plant

Pezzin M, Gonçalves R, Minozzo BR, Beltrame FL, Kitagawa R, Borges A (2017) Gastroprotective and anti-*Helicobacter pylori* effects of *Plectranthus barbatus* Andrews. 6th Brazilian Conference on Natural Products Nov. 5–8, 2017

Porfirio S, Fale PLV, Madeira PJA, Florêncio H, Ascensão L, Serralheiro MLM (2010) Antiacetylcholinesterase and antioxidant activities of *Plectranthus barbatus* tea, after in vitro gastrointestinal metabolism. Food Chem 122(1):179–187

Rajkumar K, Malathi R (2016) Anticancer activity of an ethanol extract of *Coleus forskohlii* root against gastric cancer cell lines in vitro. Int J Pharm Sci Res 7:1603–1611

Rajkumar K, Malathi R (2015) Phytochemical investigation GC-MS analysis and in vitro antimicrobial activity of *Coleus forskohlii*. Bangladesh J Pharmacol 10:924–930

Ramkumar B, Logeshwaran A, Vedhanayaki TC (2019) Phytochemical sccreening and evaluation of antimicrobial activity of *Plectranthus barbatus* rhizome. Pramana Res J 9(8):38–46

Rasineni GK, Siddavattam D, Reddy AR (2008) Free radical quenching activity and polyphenols in three species of *Coleus*. J Med Plants Res 2(10):285–291

Reddy CS, Desireddy RB, Ciddi V (2005) A review on forskolin: a cyclic AMP modulator from tissue cultures of *Coleus forskohlii*. Pharmacogn Mag 1:85–88

Rembold CM, Chen XL (1998) Mechanisms responsible for forskolin-induced relaxation of rat tail artery. Hypertension 3:872–877

Ríos-Silva M, Trujillo X, Trujillo-Hernández B, Sánchez-Pastor E, Urzúa Z, Mancilla E, Huerta M (2014) Effect of chronic administration of forskolin on glycemia and oxidative stress in rats with and without experimental diabetes. Int J Med Sci 11:448–452

Roger PP, Servais P, Dumont JE (1987) Regulation of dog thyroid epithelial cell cycle by forskolin and adenylate cyclase activator. Exp Cell Res 172:282–292

Runyoro DKB, Matee MIN, Ngassapa OD, Joseph CC, Mbwambo ZH (2006) Screening of Tanzanian medicinal plants for anti-Candida activity. BMC Complement Altern Med 6:11

Rupp RH, De Souza NJ, Dohadwalla AN (1986) Biological activities of forskolin. Proceedings of an international symposium on Forskolin: Its Chemical, Biological and Medical Potential. Hoechst India Limited, Bombay. pp.19–30

Russo I, Doronzo G, Mattiello L, De Salve A, Trovati M, Anfossi G (2004) The activity of constitutive nitric oxide synthase is increased by the pathway cAMP/cAMP-activated protein kinase in human platelets. New insights into the antiaggregating effects of cAMP-elevating agents. Thromb Res 114:265–273

Ryan DH, Bray GA, Helmcke F, Sander G, Volaufova J, Greenway F et al (1999) Serial echocardiographic and clinical evaluation of valvular regurgitation before, during, and after treatment with fenfluramine or dexfenfluramine and mazindol or phentermine. Obes Res 7:313–322

Saeed MEM, Meyer M, Hussain A, Effert T (2016) Cytotoxicity of south-African medicinal plants towards sensitive and multidrug-resistant cancer cells. J Ethnopharmacol 186:209–223. https://doi.org/10.1016/j.jep.2016.04.005

Saettone MF, Burgalassi S, Giannaccini B (2009) Preparation and evaluation in rabbits of topical solutions containing forskolin. J Ocul Pharmacol Ther 5:2. https://doi.org/10.1089/jop.1989.5.111

Saito M, Hara M, Hirotsu S (2000) Lipid decomposition promoters and cosmetics containing plant extracts. Japanese Patent JP 2000016916 A

Saklani S, Gahlot M, Kumar A, Singh R, Patial R, Kashyap P (2011) Antimicrobial activity of extracts of the medicinal plant *Coleus forskohlii*. Int J Drug Res Technol 1:52–59

Salehi B, Staniak M, Czopek K, Stepién A, Dua K, Wadhwa R, Chellappan DK, Sytar O, Brestic M, Bhat NG, Kumar NVA, Contreras MDM, Sharopov F, Cho WC, Sharifi-Rad J (2019) The therapeutic potential of the labdane diterpenoid forskolin. Appl Sci 9:4089. https://doi.org/10.3390/app9194089

Salim MLD (2003) Effect of aspirin and forskolin on blood platelet function. Alex J Pharm Sci 17: 151–155

Sano M (1983) Forskolin and adenylate cyclase. Tanpakushitsu Kakusan koso. Protein Nucleic Acid Enzyme 28(7):943–950. (in Japanese)

Sapio L, Gallo M, Illiano M, Chiosi E, Naviglio D, Spina A, Naviglio S (2017) The natural cAMP elevating compound forskolin in cancer therapy: is it time? J Cell Physiol 232:922–927

Saraswati, Katnoria JK, Nagpal AK (2016) Analytical techniques for phytochemicals screening and bioactivities of some *Coleus* species: a review. J Pharm Sci Res 8(4):227–237

Sashidhara KV, Rosiah JN, Kumar A, Bid HK, Konwar R, Chattopadhyay N (2007) Cell growth inhibitory action of an unusual labdane diterpene, 13-epi-sclareol in breast and uterine cancers in vitro. Phytother Res 21:1105–1108

Sasikala T, Prabakaran R, Sathiyapriya S (2014) Evaluation of antimicrobial activities on *Plectranthus barbatus* L. tuber. Int J Ayurv Herbal Med 4(5):1596–1601

Saunier B, Dib K, Delemer B et al (1990) Cyclic AMP regulation of Gs protein: thyrotropin and forskolin increase the quantity of stimulatory guanine nucleotide–binding proteins in cultured thyroid follicles. J Biol Chem 265:19942–19946

Schimmel RJ (1984) Stimulation of cAMP accumulation and lipolysis in hampster adipocytes with forskolin. Am J Phys 246:C63–C68

Schlepper M, Thormann J, Mitrovic V (1989) Cardiovascular effects of forskolin and phosphodiesterase-III inhibitors. Basic Res Cardiol 84:S197–S212

Schultz C, Bossolani MP, Torres LMB, Lima-Landman MTR, Lapa AJ, Souccar C (2007) Inhibition of the gastric H+, K+-ATPase by plectrinone a, a diterpenoid isolated from *Plectranthus barbatus* Andrews. J Ethnopharmacol 111:1–7

Seamon K, Daly J (1981) Forskolin: a unique diterpene activator of cyclic AMP-generating systems. J Cyclic Nucleotide Res 7:201–224

Seamon KB, Daly JW (1986) Forskolin: its biological and chemical properties. Adv Cycl Nucleotide Protein Phosphorylation Res 20:1–150

Seamon KB, Padgett W, Daly JW (1981) Forskolin: unique diterpene activator of adenylate cyclase in membranes and in intact cells. Proc Natl Acad Sci U S A 78:3363–3367

Senthilkumar CS, Suresh Kumar M, Rajasekara Pandian M (2010) In vitro antibacterial activity of crude leaf extracts from *Tecoma stans* (L) Juss. et Kunth, *Coleus forskohlii* and *Pogostemon patchouli* against human pathogenic bacteria. Int J Pharm Tech Res 2(1):438–442

Seto C, Eguchi S, Araie M, Matsumoto S, Takase M (1986) Acute effects of topical forskolin on aqueous humor dynamics in man. Jpn J Ophthalmol 30:238–244

Shaker KH, Yonus MA, Ibrahim MA, Kilany M, Wiggers F (2022) Antimicrobial activity of *Coleus forskohlii* fractions, isolation and characterization of phenolic compounds. Curr Bioact Compd 18(1):e160921190769

Shanmugam S, Pradeep BV (2019) Studies on phytochemical screening and antibacterial activity of rhizome extracts of *Coleus forskohlii* Briq. J Pure Appl Microbiol 13(3):1703–1710

Shanmugam S, Usha Rani V, Pradeep BV (2018) Antioxidant activity of rhizome extracts of *Coleus forskohlii* Briq. Asian J Pharm Clin Res 11:11. https://doi.org/10.22159/ajpcr.2018.v11i11.27125

Sharma Y, Vasundhara M (2011) Coleus (*Plectranthus barbatus*) – a multipurpose medicinal herb. Int Res J Pharm 2(3):47–58

Shivaprasad HN, Gopalakrishna S, Marianna B, Thekkoot M, Reddy R, Tippeswamy BS (2014) Effect of *Coleus forskohlii* extract on cafeteria diet-induced obesity in rats. Pharm Res 6(1):42–45

Sidhu JS, Omiecinski CJ (1996) Forskolin mediated induction of CYP3A1 mRNA expression in primary rat hepatocytes is independent of elevated intracellular cyclic AMP. J Pharmacol Exp Ther 276(1):238–245

Siegl AM, Daly JW, Smith JB (1982) Inhibition of aggregation and stimulation of cyclic AMP generation in intact human platelets by the diterpene forskolin. Mol Pharmacol 21:680–687

Singh S, Singh N (2016) Phytochemical and antimicrobial activity of various root extracts of *Coleus forskohlii* medicinal plant. Int J Res Appl Sci Eng Technol 4(2):150–155

Sirama VO, Owuor B, Amir Y, Kokwaro J, Kasima E (2019) In-vitro anthelmintic activity study of *Plectranthus barbatus* Andr. Leaves on adult *Haemonchus contortus* worms. Intern. J Pharmacol Res 9:8. https://doi.org/10.7439/ijpr.v9i8.5222

Snowden R, Harrington H, Morrill K, Jeane L, Orian M, Lopez E et al (2014) A comparison of the anti-*Staphylococcus aureus* activity of extracts form commonly used medicinal plants. J Altern Complement Med 20:372–382

Soni H, Singhai AK (2012) Recent updates on the genus *Coleus*: a review. Asian J Pharm Clin Res 5:12–17

Staudinger JL, Ding X, Lichti K (2006) Pregnane X receptor and natural products: beyond drug-drug interactions. Expert Opin Drug Metab Toxicol 2:847–857

Sun B, Geng S, Huang X, Zhu J, Liu S, Zhang Y et al (2011) Coleusin factor exerts cytotoxic activity by inducing G0/G1 cell cycle arrest and apoptosis in human gastric cancer BGC-823 cells. Cancer Lett 301:95–105

Suresh V, Deepika G, Bantal V, Beedu SR, Rupula K (2018) Evaluation of anti-inflammatory and anti-nociceptive potentials of andrographolide and forskolin: *in vivo* studies. J Biol Act Prod Nat 8(5):326–334

Tamasiro V, Davino SC, Freitas PCD, Barros SBM (1998) In vitro antioxidant activity of *Coleus barbatus* (Andr.) Benth. (false boldo) and *Peumus boldo* (Molina) (boldo do Chile): a comparative study. Rev Farm Bioquim Univ Sao Paulo 34:15–17

Tanaka R (2003) Dentifrice compositions containing taurine and *Coleus forskohlii* extract. Japanese Patent JP 2003171250 A

Tandale YN, Wagh VD (2011) Formulation and evaluation of dorzolamide hydrochloride polymeric film. Int J PharmTech Res 3:1817–1824

Tiwari N, Mishra A, Bhatt G, Chaudhary A (2014) Anti stress activity (in-vivo) of forskolin isolated from *Coleus forskohlii*. Int J Pharm Phytopharmacol Res 4(3):201–204

Tsuguyoshi A (2004) Clinical report on root extract of perilla plant (*Coleus forskohlii*) Forslean in reducing body fat. Asano Institute, Tokyo, Japan, p 2004

Tsukawaki M, Suzuki K, Suzuki R, Takagi K, Satake T (1987) Relaxant effects of forskolin on Guinea pig tracheal smooth muscle. Lung 165(4):225–237

Tung Y-C, Shih Y-A, Nagabhushanam K, Ho C-T, Chen A-C, Pan M-H (2021) *Coleus forskohlii* and *Garcinia indica* extracts attenuated lipid accumulation by regulating energy metabolism and modulating gut microbiota in obese mice. Food Res Int 142:110143

Umegaki K, Yamazaki Y, Yokotani K, Chiba T, Sato Y, Shimura F (2014) Induction of fatty liver by *Coleus forskohlii* extract through enhancement of de novo triglyceride synthesis in mice. Toxicol Rep 1:787–794

Vaden SL, Adams HR (1985) Inotropic, chronotropic and coronary vasodilator potency of forskolin. Eur J Pharmacol 118:131–137

Valdes LJ, Mislankar SG, Paul AG (1987) *Coleus barbatus (C. forskohlii)* (Lamiaceae) and the potential new drug forskolin (colenol). Econ Bot 44:474–483

Veríssimo RCSS, Lins TH, de Assis Bastos ML, de Albuquerque Sarmento P, Alvino V, Araujo MGS, Silva ALL, Araújo-Júnior JX (2013) Antimicrobial activity of *Plectranthus barbatus*

(Lamiaceae). From 5th Congress of the Brazilian Biotechnology Society (SBBIOTEC) Florianópolis, Brazil

Virgona N, Taki Y, Yamada S, Umegaki K (2013) Dietary *Coleus forskohlii* extract generates dose-related hepatotoxicity in mice. J Appl Toxicol 33:924–932

Virgona N, Yokotani K, Yamazaki Y, Shimura F, Chiba T, Taki Y, Yamada S, Shinozuka K, Murata M, Umegaki K (2012) *Coleus forskohlii* extract induces hepatic cytochrome P450 enzymes in mice. Food Chem Toxicol 50:750–755

Wagh VD (2007) Development and evaluation of ophthalmic drug delivery systems for natural forskolin and its analogues in the treatment of glaucoma. Ph. D. Dissertation, The Tamil Nadu, Dr. M.G.R. Medical University, Chennai, India

Wagh VD, Dehghan MH, Wagh KV, Salve SA, Tandale YN (2009a) Pharmacological potential and phytopharmaceutics of *Coleus forskohlii* (Makandi): a review. J GMP Ind Pharm 3:50–54

Wagh VD, Inamdar B, Wagh KV, Samanta MK (2009b) The effect of forskolin ophthalmic inserts on intraocular pressure in rabbit eyes. Int J Pharm Pharm Sci 1:146–155

Wagh VD, Pathare YS (2012) Herbal medicines and nutritional supplements used in the treatment of glaucoma: a review. Res J Pharm Biol Chem Sci 3:331–339

Wagh VD, Patil PN, Surana SJ, Wagh KV (2012) Forskolin: upcoming antiglaucoma molecule. J Postgrad Med 58:199–202

Wagh VD, Samanta MK (2010) Formulation development and evaluation of HPMC based ophthalmic inserts drug delivery system (OIDDS) of forskolin. Indian J Pharm 1:10–13

Wagh VD, Wagh KV, Pati PP, Samanta MK (2008) Formulation and evaluation of ophthalmic insert drug delivery system (OIDDS) of forskolin. Asian J Pharm 1:222–224

Wagh VD, Wagh KV, Samanta MK (2009c) Formulation and evaluation of ophthalmic insert drug delivery system of forskolin. J Pharm Res 2:12–25

Wang H, Lou C, Ma N (2019) Forskolin exerts anticancer roles in non-Hodgkin's lymphomas via regulating Axin/β-catenin signaling pathway. Cancer Manag Res 11:1685–1696

White RE, Kryman JP, El-Mowafy AM, Han G, Carrier GO (2000) cAMP-dependent vasodilators cross-activate the cGMP-dependent protein kinase to stimulate BKCa channel activity in coronary artery smooth muscle cells. Circ Res 86:897–905

Wong S, Mok W, Phaneuf S, Katz S, Salari H (1993) Forskolin inhibits platelet activating factor binding to platelet receptors independently of adenylyl cyclase activation. Eur J Pharmacol 245: 55–61

Wysham DG, Brotherton AF, Heistad DD (1986) Effects of forskolin on cerebral blood flow: implications for a role of adenylate cyclase. Stroke 17:1299–1303

Xiao C, Cheng S, Li R, Wang Y, Zeng D, Jiang H, Liang Y, Huang R, Pan H, Wu X, Fang Y, Chen C, Li X, Zhang R, Wang X, Yang Z, Yang W (2021) Isoforskolin alleviates AECOPD by improving pulmonary function and attenuating inflammation which involves downregulation of Th17/IL-17A and NF-κB/NLRP3. Front Pharmacol 30. https://doi.org/10.3389/fphar.2021.721273

Yadava JNS, Gupta S, Ahmad I, Varma N, Tandon JS (1995) Neutralization of enterotoxins of *Escherichia coli* by coleonol (forskolin) in rabbit and Guinea pig ileal loop models. Indian J Anim Sci 65:1177–1181

Yamanaka H, Oue T, Uehara S, Fukuzawa M (2010) Forskolin, a hedgehog signal inhibitor, inhibits cell proliferation and induces apoptosis in pediatric tumor cell lines. Mol Med Rep 3(1): 133–139

Yamashita A, Takashita T, Ishihara T (2004) Antiobesity agents and food, beverage and medical compositions containing bergenin and forskolin and/or astilbin. Japanese Patent JP 2004091464 A

Yang W, Li X, Chen Z, Nie L, Wang B, Shen Z (2001) Adenylate cyclase stimulation and ocular hypertension inhibition by forskolin analogs. Yanke Yangiu 19:1–4

Yao CS, Xu YL (2001) The diterpenoid quinones from *Coleus forskohlii*. Chin Chem Lett 12 (4):339–342

Yokotani K, Chiba T, Sato Y, Kubota Y, Watanabe Y, Murata M, Umegaki K (2012) Estimation of components which induce mice cytochrome P450 in *Coleus forskohlii* extract. Pharmacometrics 82:67–73

Yokotani K, Chiba T, Sato Y, Nakanishi T, Murata M, Umegaki K (2013) Influence of dietary macronutrients on induction of hepatic drug metabolizing enzymes by *Coleus forskohlii* extract in mice. J Nutr Sci Vitaminol (Tokyo) 59:37–44

You Z-P, Xiong B, Zhang Y-L, Shi L, Shi K (2018) Forskolin attenuates retinal inflammation in diabetic mice. Mol Med Rep 17:2321–2326

Yousif MH, Thulesius O (1999) Forskolin reverses tachyphylaxis to the bronchodilator effects of salbutamol: an in-vitro study on isolated Guinea-pig trachea. J Pharm Pharmacol 51:181–186

Zhang R, Kong L, Wang G, Zuo G (2005) Manufacture of dripping pills containing *Coleus forskohlii* extract. Chinese Patent CN 1682840 A

Micropropagation of *Coleus forskohlii* 6

Abstract

Forskolin is obtained from *Coleus forskohlii,* which is the only known source of this compound. The current requirements of forskolin are met through large-scale propagation, the herbal industry faces huge problems due to variations in forskolin content in the cultivated materials. The growing demand for forskolin as a potent therapeutic agent and the problems associated with obtaining consistent yields have led to the alternative sources and evaluation of various biotechnological tools for sustainable production of forskolin. This chapter gives the protocols for micropropagation of *C. forskohlii*. Details of the explants, methods of sterilization, medium components, growth hormones, their concentrations, rooting, and acclimatization methods are given.

Keywords

Micropropagation · *Coleus forskohlii* · In vitro propagation · Morphogenesis · Multiple roots · Regeneration · In vitro rooting

Abbreviations

BAP	6-Benzyl aminopurine
B5 medium	Gamborg medium
2,4-D	2,4-Dichlorophenoxy acetic acid
2,4,5-T	2,4,5-Trichlorophenoxy acetic acid
2,4,5-TP	2,4,5-Trichloropropionic acid
GA_3	Gibberellic acid
$HgCl_2$	Mercuric chloride
IAA	Indole-3-acetic acid
IBA	Indole-3-butyric acid

T. Pullaiah, *Forskolin*, https://doi.org/10.1007/978-981-19-6521-0_6

Kn	Kinetin
MS medium	Murashige and Skoog medium
NAA	α-Naphthalene acetic acid
Picloram	4-Amino-3,5,6-trichloropicolini acid
SCC	Standard Culture Conditions

6.1 Introduction

Forskolin is obtained from *Coleus forskohlii,* which is the only known source of this compound. The importance of wild *C. forskohlii* and its forskolin was realized only 4–5 decades ago. The growing demand for forskolin as a potent therapeutic agent and health supplement for glaucoma, obesity, congestive cardiomyopathy, and asthma helped entrepreneurs to work on the species aggressively and bring it to the commercial cultivation through contract farming. The current requirements of forskolin are met through large-scale propagation, but the herbal and nutraceutical industries always look for controlled chemistry and consistency in the desired content in the cultivated crops. Somehow, the cultivated *C. forskohlii* crops show a lot of variations in forskolin content because of various factors, which range from 0.1% to 0.44% (Vishwakarma et al. 1988). The growing demand for forskolin and inherent problems associated with obtaining consistent yield from cultivated sources have led to the evaluation of various biotechnological tools for sustainable production of forskolin. Petersen (1994) reviewed in vitro culture and the production of forskolin from *C. forskohlii.* Chandel and Sharma (1997) gave a detailed review on micropropagation of *Coleus forskohlii*. Similarly, Nagpal et al. (2008) reviewed micropropagation and in vitro production of secondary metabolites in various species of *Coleus*. Velmurugan et al. (2010) reviewed direct and indirect organogenesis in *C. forskohlii*. Recently, Mitra et al. (2020) have given a detailed review on the methods of micropropagation of *C. forskohlii*.

Selection of quality planting material with desired traits like high yield, disease / pest and drought resistance, chemical contents, etc. is very important. Tissue culture methods have been studied in the case of *C. forskohlii* to take care of large-scale plantations with desired uniform quality. Aspects related to explant, media constituents for shoot induction, rooting, acclimatization and in vitro conservation studies are discussed here.

6.2 Explant

Micropropagation protocol relies upon the type of explants, time of collection, and optimization of surface sterilization methods (Suryanarayana and Pai 1998). In case of *C. forskohlii*, the best response from the explants was reported when the plants were collected and inoculated from the months of March to August (Kaul et al.

2015). Explants were taken either from in vitro plants or field grown (in vivo) plants or from both sources (Thangavel et al. 2014a, b). Sen et al. (1992) used 30 days old aseptically germinated seedlings as the experimental material for shoot tip and callus culture. The shoots tips and side shoots collected from plants maintained ex situ have been the preferred choice of explants for initiating in vitro cultures by many workers (Mukherjee et al. 1996; Tefera 1998).

The choice of the explants is an important factor when developing a process for in vitro propagation. In *C. forskohlii* various explants—shoot tip, nodal segment, root, hypocotyl, and leaf—have been utilized in both direct and indirect organogenesis experiments. Most commonly, the shoot tip (Sen and Sharma 1991; Sen et al. 1992; Reddy et al. 2001; Bhattacharya and Bhattacharya 2001; Asamenew and Narayanaswamy 2004; Rajasekharan et al. 2005, 2010; Sreedevi 2006; Sreedevi et al. 2013b; Vibhuti and Kumar 2019), and nodal segment (Sharma et al. 1991; Asamenew and Narayanaswamy 2004; Rajasekharan et al. 2005, 2010; Sreedevi 2006; Dube et al. 2011; Sahai and Shahzad 2013; Sreedevi et al. 2013b, Sreedevi and Pullaiah 2014; Thangavel et al. 2014a, b; Mahmoud et al. 2019) were used in almost all micropropagation experiments. However, shoot tip is the explant of the choice since it is easily obtainable and also produces genetically stable propagules (Mitra et al. 2020). In contrast, during screening experiments by Sreedevi et al. (2013b) it was found that nodal explants exhibited relatively better response. Nodal explants implanted on MS medium with 0.5 mg/l BAP actively proliferated after 7–10 days of initiation resulting in higher percent (96.67%) of multiple shoots, whereas shoot tip exhibited a lower percentage of frequency (76.67%) with delayed response starting from 12 to 14 days. Leaf was also used as an explant source for both direct and indirect experiments (Figs. 6.1a and 6.2a, b) (Reddy et al. 2001; Ashwinkumar 2006; Krishna et al. 2010; Sahai and Shahzad 2010; Gopi and Mary 2014; Gangopadhyay et al. 2016; Vibhuti and Kumar 2019). Sreedevi et al. (2013a) used petiole and internode also as explants for callus induction and the best response was obtained with leaf explants for callus induction. Senthil Kumar et al. (2019) used young leaves as an explant source for both direct and indirect regeneration. Sen et al. (1992) used hypocotyls as explants while Tripathi et al. (1995) used root as explants. Praveena et al. (2012) and Kaul et al. (2015) reported that among the three explants (leaf, node, and shoot tip) shoot tip showed the best response.

When apical shoot tip and nodal segment were used as explant on MS media supplemented with NAA (1 mg/ml) and BAP (2 mg/ml) the explant propagated effectively after 15 days of incubation (Sharan et al. 2014). Compared to the nodal segment (as explant) the proliferation in apical tip (as explant) was more robust. Only shoot induction could be visualized when nodal segment was used as explant but, both rooting and shooting could be noticed in apical tip when used as explant. For propagation of both the explants composition of media was same (MS, supplemented with BAP (2 mg/l) and NAA (1 mg/l). The media has influenced both rooting, shooting, and multishooting in the apical tip its effect has been restricted to shooting and multishooting in the nodal segment (Sharan et al. 2014).

With a view to establishing clean cultures various surface sterilants with various concentrations and treatment durations are employed. The leaves/shoots are washed

Fig. 6.1 Direct regeneration of shoots from leaf explants of *Coleus forskohlii*: (**a**) Shoot buds regenerated on MS medium with 5.0 mg L-1 BAP after 28 days; (**b**) Shoot bud elongation on MS medium with 0.1 mg L^{-1} BAP and 0.1 mg L^{-1} IAA; (**c**) Rooting of regenerated shoot on ½ MS media; (**d**) Plants transferred in greenhouse (Source: Krishna et al. 2010)

thoroughly with Tween-20 followed by tap water and treated with 0.1% Bavistin for about 45 min. For sterilization of actively growing shoots which are most sensitive to the sterilizing process, a weak solution of $HgCl_2$ (0.1–0.2%) is generally used. Sreedevi and Pullaiah (2014) and Thangavel et al. (2014a, b) used 70% (v/v) ethyl alcohol as surface sterilant while Bhattacharya and Bhattacharya (2001) and Dube et al. (2011) used antibiotics like streptocycline as a surface sterilant. If the duration of sterilization is correctly chosen microorganisms are destroyed and plant tissues are not damaged. Treatment with 0.1% mercuric chloride solution for 4–5 min resulted in 80–95% sterile cultures (Kaul et al. 2015). If the treatment time is below 4 min or above 5 min, the sterilization percentage is low and the culture gets dehydrated. Similarly, explants treated with 70% ethyl alcohol get dehydrated, irrespective of the duration of the treatment.

Chouhan et al. (2020) also reported that the explant when treated with 0.1% mercuric chloride solution for 4–5 minutes, offered a more sterile culture. Surface sterilization with mercuric chloride with concentrations (0.1 g/L) for15 min was carried out by Mahmoud et al. (2019).

Fig. 6.2 Direct regeneration of shoots from leaf explants of *Coleus forskohlii*: (**a**) Shoot buds regenerated on MS medium with Kinetin 2.0 mg/l + NAA 0.5 mg/l after 25 days; (**b**, **c**) Multiple Shoots development on KIN 1 mg/l + NAA 0.1 mg/l; (**d**, **e**) Elongated shoots; (**f**). Rooting of regenerated shoot on ½ Strength MS media; (**g**, **h**) In vitro Accamalization and hardening; (**i**) Plants transferred in greenhouse conditions (Source: Senthil Kumar et al. 2019)

6.3 Media Constituents for Shoot Induction

MS medium (Murashige and Skoog 1962) has been used in most of the micropropagation experiments of *C. forskohlii* in both direct and indirect organogenesis. Mandler-Henger (1988), however, used BS medium (Gamborg et al. 1968) for both callus formation and shoot induction. Swaroopa et al. (2016) tested the effectiveness of three media—MS, B5, and LS—for callus induction and they reported that MS medium was more effective for callus induction. Mahmoud et al. (2019) used MS medium for micropropagation of *C. forskohlii* (Syn.: *Plectranthus barbatus*).

Three percent (w/v) sucrose has been used in most of the experiments as a carbon source (Rajasekharan et al. 2010; Sahai and Shahzad 2010). Sreedevi et al. (2013b) suggested the usage of cane sugar instead of sucrose displayed a positive response.

Shoot tips and nodal segments were cultured by Asamenew and Narayanaswamy (2004) on MS medium fortified with various concentrations of Kn. Results were better on MS medium supplemented with 1 mg/l Kn. Rajasekharan et al. (2005) reported that multiple shoots were induced from shoot tips and nodes on MS basal medium supplemented with 0.54 μM NAA and 8.87 μM BAP. When the concentration of BAP was gradually decreased, it led to amplification of shoot multiplication and finally BAP was omitted after 4 months. Rajasekharan et al. (2010) reported that *C. forskohlii* culture could be maintained at Standard Culture Conditions (SCC) for a period of 3 months and at 10 °C with low light intensity for a period of 6 months without subculture.

B5 Medium (Gamborg et al. 1968) has been used for both callus induction and for plant regeneration by Balasubramanya et al. (2012). Sreedevi et al. (2013a) reported that for callus induction, in the presence of 2,4-D (2 mg/l), B5 medium responded best for all explants when compared to MS medium, but better multiplication rate and healthy shoot buds developed on MS medium. Thus, indirect organogenesis was done in two steps.

Plant cultures are greatly influenced by physical factors such as temperature, relative humidity, and light. In *C. forskohlii*, the desirable photoperiod for better micropropagation was found to be 40–200 μmol/m^2/s. Further, the photoperiod duration was similar for all in vitro experiments, i.e., 16 h light and 8 h dark. In *C. forskohlii*, the temperature regimes for all the in vitro experiments were maintained in the range of 25 ± 2 °C (Kaul et al. 2015; Vibhuti and Kumar 2019).

High forskolin-yielding plants of *C. forskohlii* by somaclonal variation or UV mutation were made by Mandler-Henger (1988). The cultures were established on B5 medium with growth regulators, 2, 4-D and Kn, for callusing and BAP and IAA for shooting and rooting, respectively. Essentially, the same was done with micro calli after treating them with different doses of mutagenic UV.

Sen and Sharma (1991) could obtain shoot multiplication in vitro from shoot tip explants within 20–25 days from 30 days old aseptically germinated seedlings of *C. forskohlii*, using 2 mg/l of BAP. Different auxins supplemented at the level of 0.05 mg/l with BA did not yield better results. Sharma et al. (1991) reported that in vitro clonal multiplication of *C. forskohlii* has been achieved on MS medium supplemented with Kn (2 mg/l) and IAA 1 mg/l) using nodal segments. Shoots multiplied at a rate of 12-fold every 6 weeks.

The shoot cultures initiated from shoot tip explants developed multiple shoots with unorganized tissues at the cut ends on un-supplemented MS basal medium as well as on MS medium supplemented with BAP (0.5–2.5 mg/l) (Sen et al. 1992). The best effect was observed with 2 mg/l BAP. Malathy and Pai (1999) obtained callus by culturing leaf, stem, and root on MS medium supplemented with auxins NAA, 2,4-D, IBA, and IAA. Callus of stem origin exhibited a greater potential for organogenesis than callus originating from other organs.

Asamenew and Narayanaswamy (2000) could obtain regenerated plants of *C. forskohlii* in a two-step culture. Callus was obtained from shoot tip explants cultured on MS medium containing IAA (1 mg/l) and BAP (1.5 mg/l). From compact greenish callus on passage to MS basal medium containing various concentrations and combinations of IAA and Kn adventitious shoots (17.33) were obtained. On the medium containing IAA (1 mg/l) and Kn (2 mg/l) the response was better. The authors also presented a method for the rapid in vitro propagation of *C. forskohlii* from shoot tips and nodal segments which were cultured on MS medium fortified with different concentrations of Kn and better results were obtained with 1 mg/l Kn (Asamenew and Narayanaswamy 2004).

Bhattacharya and Bhattacharya (2001) reported a maximum of 12 multiple shoots from shoot tip explants of *C. forskohlii* on MS medium fortified with 0.46 mM Kn and 0.57 mM IAA. These multiple shoots exhibited 100% sprouting and developed into individual shoots.

Reddy et al. (2001) cultured leaf segments on MS medium fortified with Kn (0.4–9.3 μM) alone or in combination with 2.4-D (2.2 μM or 4.5 μM). After 4 weeks of inoculation, all the cultured leaf segments showed growth of callus from the cut ends of leaf. However, Kn alone showed prominent growth of callus. Supplementation of 2,4-D did not enhance callus formation. The callus developed on this medium subsequently turned brown within 4 weeks of culture. When this callus was subcultured on MS medium containing various amounts of Kn, the formation of adventitious shoot buds was noticed from the surface of the callus within 3 weeks of culture. Formation of leaves and shoot elongation were noticed within 8 weeks of culture. The frequency of shoot regeneration was the highest at 4.6 μM Kn.

Protocol for indirect regeneration of *C. forskohlii* from leaf explants was given by Anbazhagan et al. (2005). Callus induction was obtained on MS medium fortified with BAP (1 mg/l) along with NAA (2 mg/l). After 7 weeks of initial culture regeneration of shoots was observed. On medium fortified with 1 mg/l BA alone maximum number of shoots (23.80 $\pm$ 1.47 shoots) were obtained.

Different urea-derived herbicides and various cytokinin analogues were used to determine their effects on callusing response and shoot regeneration capacity of *C. forskohlii* (Srinivasan et al. 2006). Among them, monuron and diuron evoked profuse callusing responses from leaf segments on MS medium. Herbicide monuron (2 mg/l) showed a maximum of 3 multiple shoots/explant with a frequency of 75%

whereas diuron (0.5 mg/l) showed a higher frequency of shoot regeneration (90%) with a mean number of 6 shoots/explant.

MS medium supplemented with 5 mg/l BAP promoted regeneration of multiple shoots through direct organogenesis from the leaf (Fig. 6.1a–d) (Krishna et al. 2010). These were further elongated on MS media supplemented with 0.1 mg/l BAP and 0.1 mg/l IAA. Mahmoud et al. (2019) cultured the nodes on MS medium fortified with different concentrations of cytokinins (BA, Kn, and TDZ). The mean number of axillary shoots per explant of *C. forskohlii* reached the highest value of 6.19 on MS medium containing 2.0 mg/l TDZ. Where, the highest value of mean length was 6.44 cm on MS medium containing 1.0 mg/l Kn.

Senthil Kumar et al. (2019) obtained in vitro regeneration using leaf explants of *C. forskohlii* through indirect and direct organogenesis. In direct regeneration method, the maximum direct shoot proliferation was observed on leaf explant in Kn (2.0 mg/l) with NAA (0.1 mg/l) (Fig. 6.2a–i). In the indirect regeneration method, the growth regulator combination which produced highest percentage of organogenic callus induction from leaf explant was Kn (1.0 mg/l) (Fig. 6.3a–h). Highest shoot buds and multiple shoots were produced from callus clump with Kn (1.0 mg/l) + NAA (0.1 mg/l).

Rajasekharan et al. (2010) reported that the nodal segments of *C. forskohlii* cultured on ½MS + 8.84 μM BAP + 0.54 μM NAA produced 2 shoot buds after 8–10 days of culture. Within 90 days *C. forskohlii* achieved a height of 3.53–4.80.

The effect of various auxins, i.e., NAA, IBA, and IAA (0.1–1.0 μM) along with optimal concentration of BAP was assessed on leaf explants by Sahai and Shahzad (2010). They reported that BAP (2 μM) in combination with NAA (0.1 μM) proved to be most responsive with approximately 35 shoots/ explant and 5.4 cm mean shoot length/ culture. Histological studies by Sahai and Shahzad (2010) clearly established the endogenous origin of shoot buds from the Base Petiole Transition (BPT) region. Shoot buds can either originate endogenously from perivascular cambium or exogenously from single epidermal cells. Meristematic zone formation by continuously dividing cells in cortex near the vascular region showed the perivascular origin of shoot buds (Fig. 6.4 A-E). The formation of shoot buds from the leaf surface was found to be exogenous with only the epidermis being involved in shoot formation.

Among two cytokinins (BAP and Kn), BAP at 1.5 mg/l was found to be superior for shoot regeneration (Dube et al. 2011). Among different concentrations of BAP, 1.5 mg/l was found to be optimum in terms of percent response (96.7), shoot length (5.1 cm), and number (4.7) of shoots per explant. Balasubramanya et al. (2012) reported that the leaf explants inoculated on medium supplemented with Kn at a concentration of 3 mg/l produced hard nodular callus intermixed with small green and white protuberances. Shoot buds and roots were regenerated randomly from the surface of the callus.

Praveena et al. (2012) reported that MS medium fortified with 2,4-D at 1.5 mg/l registered highest response for callus induction. Among explants, shoot tip explants registered the highest growth (898.04 mg fr. Wt.) at 1 mg/l IBA followed by leaf lamina explants (575.85 mg fr. Wt.). 2, 4-D at various concentrations induced the highest growth increment and formation of callus in node and leaf lamina explants

Fig. 6.3 Mass Shoot propagation from the leaf-derived organogenic callus of *C. forskohlli* (**a**) leaf explant; (**b**) organogenic callus from leaf explant in KIN 1.0 mg/l; (**c**) Shoot bud induction from leaf-derived callus; (**d**) mass multiplication in Kn 1.0 mg/l + NAA 0.5 mg/l; (**e**) elongated shoots; (**f**) Rooting of regenerated shoot on ½ MS media; (**g**) In vitro Accamalization and hardening; (**h**) Plants transferred in greenhouse conditions (Source: Senthil Kumar et al. 2019)

(Table 6.1). Callus was initiated in all the explants after 2 weeks in culture. Shoot tip explants were more proliferative and registered the highest growth 1951.94 mg fr. Wt. at 1.5 mg/l 2, 4-D concentration. The authors reported that 0.5 mg/l BAP was found best for shoot induction.

Sahai and Shahzad (2013) reported a high-frequency regeneration system for micropropagation of *C. forskohlii*. Shoot multiplication was achieved through

Fig. 6.4 (**a**, **b**) Shoot induction from base petiole transition (BPT) in leaf explants on BA (2 μM), (**c**) Whole basal leaf lamina involved in shoot organogenesis on BA (2 μM) in 4-week-old culture, (**d**) Shoot multiplication and elongation in 4-week-old culture on combination medium, and (**e**) Shoot proliferation in 6-week-old culture on BA (2 μM) + NAA (0.1 μM) (Source: Sahai and Shahzad 2010)

axillary bud development and direct adventitious shoot formation in nodal explants on MS medium containing BAP (5 μM).

For induction of callus Atulkar et al. (2014) inoculated the leaf explants on MS medium fortified with various concentrations of 2,4-D ranging from 0.01 to 2.5 mg/l. The callus was subcultured onto MS medium fortified with various concentrations and a combination of auxins (NAA) and cytokinin (BAP) showed an organic response and produced multiple shoots.

Gopi and Mary (2014) reported that highest callus induction was 80% at the end of the fourth week and it was recorded on a medium fortified with 0.5 mg/l of 2,4-D and 1.0 mg/l of BAP. Somatic embryogenesis was achieved on medium fortified with the combination of BAP (1.0 mg/l) + NAA (1.0 mg/l) + Kn (0.5 mg/l). The highest number of somatic embryos per culture was 28.5.

Sreedevi and Pullaiah (2014) described a large-scale micropropagation protocol for *C. forskohlii*. MS medium fortified with BAP (0.25 mg/l) and Kn (0.25 mg/l) resulted in an average of 41 multiple shoots. Thangavel et al. (2014a, b) reported a maximum number of 9.95 shoots with 88.6% regeneration frequency from in vitro nodal explants cultured on MS medium fortified with 4.65 μM Kn. In vivo node explants produced a maximum of 8.47 shoots on MS medium fortified with 4.65 μM Kn + 1.73 μM GA_3. Shoot apices when cultured on a medium with 2 mg/l BAP showed the best morphogenetic response (with 15–20 shoots per explants and 80% response) (Kaul et al. 2015).

Vibhuti and Kumar (2019) cultured shoot tip, node, and leaf on MS medium with different plant growth regulators. Adventitious shoots developed after 7 days from shoot tip explants when cultures on a medium containing 2 mg/ BAP and 1 mg/l Kn while callus was formed from leaf explants on a medium fortified with 1 mg/l NAA.

Table 6.1 A detailed report on various tissue culture studies of *Coleus forskohlii* to date

Explant/s	Culture medium + Plant Growth Regulators (PGRs) mg/l	Response	References
Leaf	For callus induction B5 + 2.3 μM 2,4-D + 0.93 μM Kn For shoot induction B5 + 5μMBAP	Callus formation Shoot induction	Mandler-Henger 1988
Shoot tip	MS+ 2 mg/l BAP	Multiple shoots	Sen and Sharma 1991
Nodal segment	MS + 2 mg/l Kn + 1 mg/l IAA For rooting MS + 1 mg/l IAA	Multiple shoots Rooting	Sharma et al. 1991
Shoot tip	MS + 2 mg/l BAP	80–90% response	Sen et al. 1992
Leaf, stem, root	MS+ combination of different concentrations of IAA + IBA		Malathy and Pai 1999
Shoot tip	For callus induction MS + 1 mg/l IAA + 1.5 mg/l BAP For shoot induction MS + 1 mg/l IAA + 2 mg/l Kn	Callus formation and adventitious shoots	Asamenew and Narayanaswamy 2000
Shoot tip Nodal segment	MS +0.46 mM Kn +0.57 mM IAA MS +0.46 mM Kn +0.57 mM IAA	Twelve shoots per explant 100% response	Bhattacharya and Bhattacharya 2001
Leaf	For callus induction MS + 2.4 μM Kn For shoot induction MS + 4.6 μM + 0.54 μM NAA	Callus growth prominent Multiple shoot	Reddy et al. 2001
Shoot tip, nodal segment	MS+ 1 mg/l Kn	Multiple shoots	Asamenew and Narayanaswamy 2004
Leaf	For callus induction MS + 1 mg/l BAP + 2 mg/l NAA For regeneration MS + 1 mg/l BAP	Profuse callus 23 shoots	Anbazhagan et al. 2005
Shoot tip, node	MS + 0.54 μM NAA + 8.87 μM BAP	Multiple shoots	Rajasekharan et al. 2005
Nodal segment, leaf	For callus induction MS+ 2 mg/l 2,4-D For shoot induction MS + 0.5 mg/l Kn	Nodular, compact, green callus Multiple shoots	Ashwinkumar 2006
Leaf	MS+ monuron+diuron MS+ 2 mg/l monuron MS+ 0.5 mg/l diuron	Callus formation 75% regeneration 90% regeneration	Srinivasan et al. 2006
Leaf	MS + 5 mg/l BAP	76.4% response	Krishna et al. 2010
Shoot tip, nodal segment	½ MS + 8.87 μM BAP + 0.54 μM NAA	2 shoot buds per explant	Rajasekharan et al. 2010
Leaf	MS+ 2 μM BAP + 0.1μMNAA	35 shots per explant	Sahai and Shahzad 2010

(continued)

Table 6.1 (continued)

Explant/s	Culture medium + Plant Growth Regulators (PGRs) mg/l	Response	References
Nodal segment	For shoot induction MS + 1.5BAP For root induction MS+ no growth regulators	96.7% response, 4.7 shots/explant	Dube et al. 2011
Leaf	For callogenesis B5 + 2,4-D, 2,4,5-T, 2,4,5-TP, Picloram For multiple shoot induction B5+ 0.5 mg/l BAP, B5 + 0.5 mg/l Kn	Friable callus Green shoot buds without callus	Balasubramanya et al. 2012
Leaf, node, shoot tip	For callus induction MS +1.5 mg/l 2,4-D For shoot induction MS + 0.5 mg.l BAP For rooting	Shoot tip showed best response	Praveena et al. 2012
Leaf	MS + 2 μM BAP + 0.1 μM NAA	35 shoots per explants	Sahai and Shahzad 2013
Leaf, internode, petiole, leaf	For callus induction B5 + 2 mg/l 2,4-D and MS + 2 mg/l 2,4-D For shoot induction MS+ 2 mg/l BAP + 1 mg/l NAA For rooting ½ MS	98% response was in leaf explants on B5 medium 53.33% response with leaf explant	Sreedevi et al. 2013a
Node Shoot tip	MS + 0.5 mg/l BAP MS + 0.5 mg/l BAP For rooting ½ MS	96.67% response with 36 shoots 76.67% response	Sreedevi et al. 2013b
Leaf	Callus induction MS+ 1 mg/l 2,4-D + 0.5 BAP Shoot induction MS+ 0.5 mg/l BAP + 0.1 NAA	Profuse callus 80% response with 3 shoots/explant	Atulkar et al. 2014
Leaf	For callus induction MS + 1 mg/l 2,4-D For somatic embryogenesis 1 mg/l BAP + 1 mg/l NAA + 0. mg/l Kn	80% response 72% response	Gopi and Mary 2014
Nodal segment	MS + 0.25 mg/l BAP + 0.25 mg/l Kn For rooting ½ MS	41 multiple shoots per explant	Sreedevi and Pullaiah 2014
Nodal segment	For shoot induction MS + 4.65 mM Kn For rooting MS + 7.38 μM IBA	88.6% response with 9.95 shoots per explant	Thangavel et al. 2014a, 2014b
Apical and axillary meristem	For shoot induction MS + 2 mg/l BAP	80% response	Kaul et al. 2015
Leaf			Gangopadhyay et al. 2016
Leaf	For callus induction MS+ 2 mg/l 2,4-D; MS+ 2 mg/l NAA	76.7% response	Swaroopa et al. 2016

(continued)

Table 6.1 (continued)

Explant/s	Culture medium + Plant Growth Regulators (PGRs) mg/l	Response	References
Node	MS + 2 mg/l TDZ		Mahmoud et al. 2019
Shoot tip leaf node	MS+ 1.5 mg/l BAP MS+ 1 mg/l Kn MS+ 1.5 mg/l BAP	Multiple shoots Callus growth Multiple shoots	Vibhuti and Kumar 2019
Nodal	For shoot induction MS+ 4.44 μM BAP For rooting ½ MS+ 2.46 μM IBA	80% response, 24.3 shoots per explants 100% response	Janarthanam and Sumathi 2020

Shoots originated directly at the axillary position of the nodal explants in presence of cytokinins in the medium. Among two cytokinins (BAP and Kn), BAP at 1.5 mg/l was found to be superior and production of multiple shoots. Janarthanam and Sumathi (2020) reported that after 30 days of culture 24.3 shoots/ explant developed on MS medium fortified with 4.44 μM BAP with 80% response and an average length of 5.6 cm (Fig. 6.5a–f).

Many growth adjuvants have been used for both callus induction and shoot multiplication and coconut water is one among them. The main advantage of using coconut water is to replace the hormones and to reduce the cost of the media. In *C. forskohlii*, Sreedevi et al. (2013b) reported that MS basal medium with coconut water exhibited multiple shoot regeneration, but lower than the medium fortified with BAP (0.5 mg/l).

Different explants like leaf, node, and shoot tip of *C. forskohlii* were cultured by Chouhan et al. (2020) on different hormone concentrations of MS medium. Nodal explants in MS medium containing 2.0 mg/l BAP showed better multiplication at the rate of 4–8 shoots per explant. Callus from leaves of *C. forskohlii* has been induced in MS medium containing 1 mg/l 2,4-D and 0.5 mg/l BAP after 5 weeks of incubation. The explants were subcultured and more shoots were obtained using the growth hormones 1 mg/l IAA and 2 mg/l Kn. Chouhan et al. (2020) also studied gene expression in *C. forskohlii* through plant tissue culture. In vitro optimal multiplication of 20.4 ± 0.28 shoots per explant was achieved by Chakraborty et al. (2022) in MS medium fortified with 1.0 mg/l BAP and 0.1 mg/l NAA.

6.4 Rooting and Acclimatization

Sen and Sharma (1991) reported that in 90% of the excised shoots, rooting occurred on an un-supplemented basal medium within 20 d. The rooted shoots transferred to soil showed 60% survival. Rooting was achieved by Sen et al. (1992) upon transfer of shoots onto MS medium containing IAA (1 mg/l). In vitro produced plants were established in soil with almost 100% survival.

Fig. 6.5 (**a**) Mother plant of *Coleus forskohlii* (**b–f**). Stages of in vitro plant regeneration from nodal explants of *Coleus forskohlii.* (**b**). Initiation of multiple shoots from nodal explant inoculated on MS medium supplemented with BAP 4.44 μM after 2 weeks of culture. (**c**). Proliferation of multiple shoots from nodal explants on MS containing BAP 4.44 μM after 30 days of culture. (**d**). Rooting of in vitro regenerated shoots inoculated on half-strength MS containing 2.46 μM IBA. (**e**) Healthy in vitro developed plantlet. (**f**) Well established and hardened in vitro plants transferred to small paper pots (Source: Janarthanam and Sumathi 2020)

In general, rooting of regenerated shoots is a separate step in many micropropagation protocols. However, the total time schedule was condensed to only one step by Bhattacharya and Bhattacharya (2001). In vitro produced plants were established in soil with almost 100% survival. Velmurugan (2007) reported that for better rooting of callus-derived microshoots, the ideal medium was found to be half strength MS medium fortified with IBA (0.2 mg/l).

Reddy et al. (2001) and Dube et al. (2011) also reported that rooting occurred spontaneously in the regenerated on half-strength MS medium without growth regulators. Leafy shoots derived from leaf callus were transferred to MS basal medium for rooting and 1 week after inoculation, root formation was noticed from the basal cut portion of the shoot. Reduction of MS salts to half-strength enhanced root formation from the shootlets. The maximum frequency of root formation (80%), number (25), and length (7.5 cm) of the roots were achieved in half-strength MS medium without growth regulators. Rajasekharan et al. (2005, 2010) also reported that rooting was simultaneous and occurred in the same medium on which multiple shoots were induced, and there was no need to use a separate medium for rooting. Rooting in multiplication media enabled successful establishment ex vitro and 85% survival. High rooting frequencies were obtained from shoots of in vitro grown plantlets by Sreedevi et al. (2013a, b) and Sreedevi and Pullaiah (2014) ½ MS devoid of plant growth regulators. Regenerated plants were acclimatized in greenhouse conditions and then transferred to soil with 100% survival.

Krishna et al. (2010) transferred regenerated and elongated shoots to half-strength MS medium with 1.5% sucrose which resulted in profuse rooting. Rooted plants were acclimatized and transferred to soil and were maintained in a greenhouse.

Regenerated microshoots were transferred by Sahai and Shahzad (2010) to a root induction medium composed of half-strength MS basal medium devoid of any PGR as well as in combination with varying concentrations of IBA, NAA, and IAA (1–5 μM) individually. Rhizogenesis was induced in the 2-week old culture. Regenerated microshoots rooted spontaneously in half-strength MS medium devoid of PGR. Rooted plantlets were transferred to thermocol cups containing soilrite for proper hardening. Thereafter, the plants were potted in earthen pots containing a mixture of garden manure: garden soil: soilrite (1: 2: 1) and kept under the sun in field conditions. Rooting was induced by Praveena et al. (2012) with nodal explants, cultured in MS medium fortified with BAP 1.5 mg/l + 2, 4-D 2 mg/l.

Regenerated shoots were subcultured by Thangavel et al. (2014a, b) on half-strength MS medium with the supplementation of GA_3 for shoot elongation. GA_3 + Kn (1.73+ 2.32 μM) was found to be more potent for shoot elongation (8.36 cm shoot length) followed by GA_3 (1.73 μM) alone with 7.58 cm shoot length. Micropropagated shoots were rooted by Thangavel et al. (2014a, b) on MS medium supplemented with 7.38 μM IBA, within 20 days of culture. In vitro raised plantlets were hardened and subsequently transferred to the field conditions with a 90% survival rate.

In *C. forskohlii* regeneration of shoots and roots are concomitant and no separate treatments are required (Kaul et al. 2015). Rooted plants were separated carefully and transferred to small pots containing vermiculite. Plantlets were then placed in the greenhouse for 5 days and then transferred to the field for hardening with a 100% survival rate. Senthil Kumar et al. (2019) obtained rooting of the regenerated shoots on half-strength MS medium without any growth regulators. They reported that addition of IAA and IBA suppressed rooting response of the regenerated shoots. This is in contrast to the other reports. Mahmoud et al. (2019) reported that the mean number of roots/explant of *C. forskohlii* reached the highest value and the mean lengths were 30.00 and 11.8 cm, respectively, on 1/2 MS medium containing 0.5 mg/l IBA.

Janarthanam and Sumathi (2020) reported that the in vitro shoots showed a higher response for the development of rooting on half-strength MS supplemented with 2.46 μM IBA. Maria et al. (2022) described a simple protocol for in vitro root induction of *C. forskohlii*.

The plants were successfully ex vitro adapted with 93.3% survival on the mixture substrate of soil, peat, and sand (2:1:1 v/v/v). The maximum survival of in vitro raised plantlets was obtained by transferring them to the potting mixture containing equal parts of sand: red soil: vermiculite (1:1:1). The ideal hardening environment was covering the plantlets with punched polythene covers and keeping them under a mist chamber condition exerted greater survival percentage of plantlets (Velmurugan 2007; Velmurugan et al. 2008).

Mahmoud et al. (2019) reported that the survived acclimatized plants were 93% after 2 weeks from transferring rooted plantlets on a mixture of soil, vermiculite, and sand (2:1:1: v/v/v). After a month of transferring the rooted plantlets, the percentage of surviving was 83%.

6.5 In Vitro Conservation

In vitro conservation studies of *C. forskohlii* were carried out by Rajasekharan et al. (2005) with 2–3 week maintained in vitro plants under standard and reduced culture conditions (SCC, RCC). In vitro plants could be successfully conserved in full strength MS medium (FMS) under SCC for 6 months without subculture with full potential to regenerate, producing viable shoots.

6.6 Conclusion and Future Prospects

Commensurate with the in vitro propagation studies, field performance studies of in vitro raised plants need to be explored. Unfortunately, in the case of *C. forskohlii*, very little importance is given lab to land aspect and hence restricted utility of this technology. Further, not much work has been done on the use of elicitors, genetic transformation, and somaclonal variations to develop high-yielding varieties. Translation of lab studies in the field for commercial production of desired high-yielding genotypes in terms of biomass and forskolin content and disease and pest-resistant varieties need to be looked into. The micropropagation protocols given above will be useful for regeneration of genetically modified crop varieties. Since forskolin is obtained from roots, hairy root culture using reactors needs to be studied for commercial production of the forskolin.

References

Anbazhagan K, Kumar NS, Hemavthi V, Sathyanarayana BN (2005) In vitro morphogenesis in *Coleus forskohlii*. J Med Aromatic Plant Sci 27:253–256

Asamenew MT, Narayanaswamy P (2000) Induction of callus and plant regeneration in *Coleus forskohlii* Briq. J Appl Horticulture 2:25–27

Asamenew T, Narayanaswamy MP (2004) In vitro shoot multiplication and rooting in *Coleus forskohlii* Briq. Crop Science Society of Ethiopia F01:126–134

Ashwinkumar SK (2006) Complementary and alternative systems of medicine. In: International conference on globalization of traditional. Tamil Nadu Agricultural University, Coimbatore, pp 29–30

Atulkar P, Thakur R, Singh P (2014) Shoot regeneration from leaf derived callus of *Coleus forskohlii* Briq. Int J Sci Res 3(11):2662–2664

Balasubramanya S, Rajanna L, Anuradha M (2012) Effect of plant growth regulators on morphogenesis and forskolin production in *Plectranthus barbatus* Andrews. In Vitro Cell Dev Bio-Plant 48:208–215

Bhattacharya R, Bhattacharya S (2001) In vitro multiplication of *Coleus forskohlii* Briq. an approach towards shortening the protocol. In Vitro Cell Dev Biol 37:572–575

Chakraborty A, Haque SM, Dey D, Mukherjee S, Ghosh B (2022) Detection of UTI pathogen-killing properties of *Coleus forskohlii* from tissue cultured in vitro and ex vitro plants. Proc Natl Acad Sci India Sect B Biol Sci 92:157–169. https://doi.org/10.1007/s40011-021-01285-4

Chandel KPS, Sharma N (1997) Micropropagation of *Coleus forskohlii* (Willd.) Briq. In: Bajaj YPS (ed) Biotechnology in Agriculture and Forestry, vol 40. Springer, pp 74–84

Chouhan S, Patel T, Saraf M (2020) Gene expression studies of *Coleus forskohlii* through plant tissue culture. Int J Multidiscipl Educ Res 9(7):166–175

Dube P, Gangopadhyay M, Dewanjee S, Ali MN (2011) Establishment of a rapid multiplication protocol of *Coleus forskohlii* Briq. and in vitro conservation by reduced growth. Indian J Biotechnol 10:228–231

Gamborg OL, Miller RA, Ojman K (1968) Nutrient requirements of suspension cultures of soybean root cells. Exp Cell Res 50:151–158

Gangopadhyay M, Gantait S, Palchoudhury S, Ali MN, Mondal C, Pal AK (2016) UVC-priming mediated modulation of forskolin biosynthesis key genes against Macrophomina root rot of *Coleus forskohlii* - a tissue culture based sustainable approach. Phytochem Lett 17:36–44

Gopi C, Mary MR (2014) In vitro plant regeneration through somatic embryogenesis in medicinally important leaf explants of *Coleus forskohlii* Briq. J Agric Vet sci 9:20–23

Janarthanam B, Sumathi E (2020) *In vitro* plant regeneration from nodal explants of *Coleus forskohlii* Briq. - an important medicinal plant. Plant Tissue Cult Biotech 30(1):143–148

Kaul T, Malik MA, Yaqoob U, Mehta J (2015) High frequency and rapid in vitro plant regeneration of *Coleus forskohlii* Briq. Med Arom Plants 4. https://doi.org/10.4172/2167-0412.1000193

Krishna G, Reddy PS, Nair NA, Ramteke PW, Bhattacharya PS (2010) In vitro direct shoot regeneration from proximal, middle and distal segment of *Coleus forskohlii* leaf explants. Physiol Mol Biol Plants 16:195–200

Mahmoud DSM, Sayed LM, Diab MI, Fahmy EM (2019) In vitro propagation of *Plectranthus barbatus* Andrews as important medicinal plant. 14 th Conf. Agric. Develop. Res., Fac. of Agric., Ain Shams Univ., March, 2019, Cairo, Egypt Special Issue, 27(1): 511–517

Malathy S, Pai JS (1999) Monitoring of forskolin production from roots and callus by HPTLC in *Coleus forskohlii* Briq. J Spices Arom Crop 8:153–157

Mandler-Henger A (1988) Somaclonal variation be *Coleus forskohlii*. Doctoral thesis. University of Tubingen

Maria G, Stancheva I, Kirova E, Petrova M, Hendawy S, Zayova E (2022) Assessment of the antioxidant power of in vitro obtained *Coleus forskohlii* Briq. J Microbiol Biotechnol Food Sci 11(4):e3840

Mitra M, Gantait S, Mandal N (2020) *Coleus forskohlii*: advancements and prospects of in vitro biotechnology. Appl Microbiol Biotechnol 104(6):2359–2371. https://doi.org/10.1007/s00253-020-10377-6

Mukherjee S, Ghosh B, Jha S (1996) Forskolin synthesis in in vitro cultures of *Coleus forskohlii* Briq. Transformed with *Agrobacterium tumefaciens*. Plant Cell Rep 15:691–694

Murashige T, Skoog F (1962) A revised medium for rapid growth and bioassays with tobacco tissue cultures. Physiol Plant 15:473–497

Nagpal A, Singh B, Sharma S, Rani G, Virk GS (2008) *Coleus* spp.: micropropagation and in vitro production of secondary metabolites. Medicinal and Aromatic Plant Science and Biotechnology 2:1–17

Petersen M (1994) *Coleus* spp.: in vitro culture and the production of forskolin and rosmarinic acid. Med Aroma Plants 4:69–92

Praveena R, Pandian AS, Jegadeesan M (2012) In vitro culture studies on medicinal herb - *Coleus forskohlii* Briq. Libyan Agric Res Center J Intern 3:30–35

Rajasekharan PE, Ambika SR, Ganeshan S (2005) In vitro conservation of *Coleus forskohlii* - an endangered medicinal plant. J Plant Biotech 7(2):1–7

Rajasekharan PE, Ganeshan S, Bhaskaran S (2010) In vitro regeneration and conservation of three *Coleus* species. Med Aromat Plant Sci Biotechnol 4:24–27

Reddy PS, Rodrigues R, Rajasekharan R (2001) Shoot organogenesis and mass propagation of *Coleus forskohlii* from leaf derived callus. Plant Cell Tissue Organ Cult 66:183–188

Sahai A, Shahzad A (2010) In vitro clonal propagation of *Coleus forskohlii* via direct shoot organogenesis from selected leaf explants. J Plant Biochem Biotechnol 19:223–228

Sahai A, Shahzad A (2013) High frequency in vitro regeneration system for conservation of *Coleus forskohlii*: a threatened medicinal herb. Acta Physiol Plant 35:473–481

Sen J, Sharma AK (1991) *In vitro* propagation of *Coleus forskohlii* Briq. for forskolin synthesis. Plant Cell Rep 9:696–698

Sen J, Sharma AK, Sahu NP, Mahato SB (1992) Production of forskolin in in vitro cultures of *Coleus forskohlii*. Planta Med 58:324–327

Senthil Kumar P, Vasuki A, Lakshmi Prabha (2019) A mass propagation of important medicinal plant *Coleus forskohlii* (Briq) through direct and indirect regeneration using leaf explant. J Emerg Technol Innov Res 6(3):178–186

Sharan AK, Singh BP, Dubey SR, Kumar R, Kishor A, Kumar G, Kumari S (2014) Effective propagation and evaluation of salt tolerance in *Coleus forskohlii*, an endangered herb. Intern J Adv Sci Eng Technol 3(2):24–31

Sharma N, Chandel KPS, Srivastava VK (1991) In vitro propagation of *Coleus forskohlii* Briq. A threatened medicinal plant. Plant Cell Rep 10:67–70

Sreedevi E (2006) Selection and in vitro multiplication of high yielding clones of *Plectranthus barbatus* Andr. [Syn.: *Coleus forskohlii* (Willd.) Briq.]. PhD thesis, Sri Krishnadevaraya University, Anantapur, India

Sreedevi E, Anuradha M, Pullaiah T (2013a) Plant regeneration from leaf-derived callus in *Plectranthus barbatus* Andr. [Syn.: *Coleus forskohlii* (Willd.) Briq.]. Afr J Biotech 12(18): 2441–2448

Sreedevi E, Anuradha M, Pullaiah T (2013b) Relative benefits of using different sugars and coconut water for *in vitro* regeneration of *Plectranthus barbatus* [Syn: *Coleus forskohlii* (Willd.) Briq.]. J Trop Med Plants 14:23–29

Sreedevi E, Pullaiah T (2014) Effect of growth regulators on *in vitro* organogenesis and long term storage of *Plectranthus barbatus* Andr. [Syn.: *Coleus forskohlii* (Wild.) Briq.]. Curr Trends Biotechnol Pharm 8(2):143–151

Srinivasan M, Nachiappan V, Rajasekharan R (2006) Potential application of urea-derived herbicides as cytokinins in plant tissue culture. J Biosci 31:599–605

Suryanarayana M, Pai JS (1998) Studies in micropropagation of *Coleus forskohlii*. J Med Aromatic Plant Sci 20:379–382

Swaroopa G, Anuradha M, Pullaiah T (2016) Influence of sucrose and plant growth regulators on growth and forskolin production in callus cultures of *Coleus forskohlii*. Curr Trend Biotechnol Pharm 10:45–54

Tefera M (1998) Callus culture and plant regeneration in *Coleus forskohlii*. MSc Thesis. University of Agricultural Sciences, Bangalore

Thangavel P, Britto SJ, Prabhu S, Senthil Lumar S (2014a) *In vitro* callus induction and plantlets regeneration from leaf explants of *Plectranthus barbatus* Andrews – a valuable medicinal plant. Int J Adv Sci Eng 1(2):51–57

Thangavel P, Prabhu S, Britto SJ (2014b) High frequency shoots regeneration from nodal explants of *Plectranthus barbatus* Andrews belong to the Lamiaceaae. J Andaman Sci Assoc 19:126–135

Tripathi CKM, Basu SK, Jain S, Tandon JS (1995) Production of coleonol (forskolin) by root callus cells of plant *Coleus forskohlii*. Biotchnology Letters 17:423–426

Velmurugan M (2007) Induced mutagenesis for improvement of yield and forskolin content in coleus (*Coleus forskohlii* Briq.). PhD thesis. Tamil Nadu Agricultural University

Velmurugan M, Rajamani K, Paramaguru P, Gnanam R, Kannan Bapu JR, Davamani V, Selvakumar T (2010) Direct and indirect organogenesis in medicinal coleus (*Coleus forskohlii* Briq.) – a review. Agric Rev 31:73–76

Velmurugan M, Rajamani K, Paramaguru P, Kannan Bapu JR, Gnanam R (2008) Induction of variation in *Coleus forskohlii* Briq. Through in vitro mutation technique. In: International Symposium on induced mutations in plants. Book of Abstracts, Vienna

Vibhuti RK, Kumar D (2019) Effect of 6-BAP on callus culture and shoot multiplication of *Coleus forskohlii* (syn *Plectranthus forskohlli* Wild) Briq. Res J Life Sci Bioinform Chem Sci 5:574–581

Vishwakarma RA, Tyagi BR, Ahmed B, Hussain A (1988) Variation in forskolin content in the roots of *Coleus forskohlii*. Planta Med 54:471–472

In Vitro Production of Forskolin

7

Abstract

Root of *Coleus forskohlii* is the only source of the labdane diterpenoid forskolin. Forskolin is useful in preventing the clotting of platelets, in reducing intraocular pressure in cases of glaucoma, as a hypotensive and as an aid to nerve regeneration following trauma. To meet the ever-increasing demand for forskolin from the pharmaceutical industry several attempts have been made to standardize and use plant tissue culture techniques. In this chapter, methods of obtaining forskolin through hairy root cultures, callus cultures, and suspension cultures have been described in detail. The effect of various biotic and abiotic elicitors on forskolin production has also been discussed.

Keywords

Forskolin · *Coleus forskohlii* · Callus cultures · Suspension cultures · Hairy root cultures

Abbreviations

2,4-D	2,4-Dichlorophenoxy acetic acid
BAP	6-Benzyl aminopurine
DDF	1,9-dideoxyforskolin
IAA	Indole-3-acetic acid
IBA	Indole-3-butyric acid
Kn	Kinetin
NAA	α-Naphthalene acetic acid
PGRs	Plant growth regulators

T. Pullaiah, *Forskolin*, https://doi.org/10.1007/978-981-19-6521-0_7

7.1 Introduction

Secondary metabolites are metabolically expensive to produce and accumulate and are therefore frequently present in plants in much smaller quantities than primary metabolites. In addition, secondary metabolites tend to be biosynthesized in specialized cell types and at distinct development stages, making their extraction, isolation, and purification difficult. Hence, secondary metabolites that are used commercially are generally higher value-low volume products. Thus, secondary metabolites can be considered special materials or fine chemicals (Balandrin and Klocke 1988).

Large-scale culturing of plant cells is carried out in cell suspension systems from which secondary metabolites can be extracted. The main advantage is it may provide a continuous, reliable source of natural products year-round and may be more easily purified, thus possibly reducing production and processing costs (Balandrin and Klocke 1988). In the intact plant, the biosynthesis of secondary metabolites thought to be protective compounds can be induced by stresses such as fungal infection.

Forskolin, a naturally occurring labdane diterpenoid, derived from the root of *C. forskohlii* is gaining its importance globally. To meet the increasing demands of pharmaceutical industry for production of forskolin attempts have been made to standardize and use plant tissue culture techniques. Petersen (1994) reviewed in vitro culture and the production of forskolin from *C. forskohlii.* Nagpal et al. (2008) reviewed micropropagation and in vitro production of secondary metabolites in various species of *Coleus*.

7.2 Organized Cultures

7.2.1 Hairy Root Cultures

Krombholz et al. (1990, 1992) established hairy root cultures using *Agrobacterium rhizogenes* strain 15,834 on Gamborg's B5 medium containing 1 mg/l IBA. The biomass multiplied 20 times in 3 weeks. Though the forskolin yield was nearly the same as with untransformed cultures, yet due to better growth, overall production of forskolin was much higher (4.5 mg/l) for transformed root cultures. Hairy roots of *C. forskohlii* were induced by Sasaki et al. (1998) by infection with the *A. rhizogenes* MAFF 03-01724 strain for high forskolin production.

Mukherjee et al. (2000b) reported enhanced forskolin production in genetically transformed cultures of *C. forskohlii* by casein hydrolysate. Mukherjee et al. (2002) reviewed the studies on genetic transformation of *C. forskohlii* for forskolin production.

The growth and rosemarinic acid (RA) production by *C. forskohlii* hairy root cultures in various liquid media were examined by Li et al. (2005). The hairy root cultures showed good growth in hormone-free MS medium containing 3% sucrose and Gamborg B5 medium containing 2% sucrose. RA yield reached 4.0 mg

(MS medium) and 4.4 mg (B5 medium) after 5 weeks of culture in a 100-ml flask containing 20 ml of each medium.

Hairy root cultures developed from leaf explants of *C. forskohlii* upon infection with *A. rhizogenes* strains A4 and LBA9402 for the production of forskolin were investigated by Kukreja and Garg (2007). The hairy roots were excised and grown on B5 medium. For root growth and biomass production B5 medium was less responsive. Hence, all hairy root lines were transferred to a hormone-free WP liquid medium which supported maximum hairy root biomass (6.33 g). The root biomass further increased by 50% reaching 9.33 g with the addition of IBA (1.0 mg/l). Forskolin content ranged from 0.011 to 0.11% for different lines of A4 strain and from 0.005% to 0.083% for LBA9402. Similarly, the level of immediate precursor, 9-deoxyforskolin was also higher for lines of A4 strain.

An experiment was conducted by Uma Maheswari et al. (2011) for induction and establishment of *Rhizobium rhizogenes* (Syn.: *Agrobacterium rhizogenes*) mediated hairy root culture of *C. forskohlii* on hormone-free semisolid MS Medium with B5vitamins. Hairy roots were induced with the strains of *Agrobacterium rhizogenes* ATCC18534. It resulted in the emergence of hairy roots from the leaf and stem explants after 20th day of infection (Fig. 7.1a, b).

Reddy et al. (2012) elucidated the effect of precursors and elicitors on the production of forskolin from transformed root cultures of *C. forskohlii.* By infecting leaf explants with *A. rhizogenes* strain A4 hairy root cultures are established on MS basal medium. Suspension cultures of hairy roots were initiated on MS medium containing 1 mg/l IBA and 600 mg/l casein hydrolysate. They investigated the biomass growth and forskolin production in suspension cultures of hairy roots. They found that forskolin production was parallel to the growth of biomass. After 5 weeks of growth maximum forskolin production was observed. Abiotic elicitors like salicylic acid, copper sulfate, methyl jasmonate, and precursors α-ketoglutaric acid and L-phenylalanine were added to hairy root cultures on different days of incubation period to increase the yield of forskolin. Addition of elicitor methyl jasmonate (500 μM) and the precursor L-phenylalanine (1 mM) on day 14 significantly enhanced the production of forskolin over the control hairy root cultures. A detailed phytochemical investigation of the secondary metabolites of *Coleus forskohlii* hairy root cultures was undertaken by Asada et al. (2012) which resulted in the isolation of 22 compounds, including four forskolin derivatives and a monoterpene.

Pandey et al. (2014) induced hairy roots from the nodal segments and mature leaf of *C. forskohlii* using *A. rhizogenes* strain MTCC 2364. The response of nodal segment was much better and produce large amount of hairy roots (Fig.7.2). Hairy roots emerged from the base of the node within 12 days of culture. When the shoot portion was cut and transferred to a new MS basal medium it started producing roots within 5 days. Forskolin content was highest in hairy roots when compared with other plant parts. Tips of hairy roots were cut down after 4 weeks of culture and inoculated onto the MS and B5 liquid and semisolid medium with cefotaxime 400 mg/l and devoid of any growth regulators. The cultures were maintained on rotary shaker at 100 rpm at 25 °C in the dark. The cultured tips showed growth

A

B

Fig. 7.1 Hairy roots induced by *Agrobacterium rhizogenes* in *C. forskohlii* leaf explants (Source: Uma Maheswari et al. 2011)

within 3 weeks of culture and these actively grown root tips were cut and subcultured at regular intervals of 3 weeks. A total of four subcultures were done. In every subculture, the concentration of cefotaxime was reduced to 200 mg/l. The hairy roots were subcultured on MS and B5 liquid media for further proliferation. The

Fig. 7.2 Hairy root culture in *Coleus forskohlii:* (**a**) After co-cultivation root emerged from explant; (**b**) After co-cultivation from nodal part of stem shoot started emerging; (**c**) After 25 days it produced lots of roots; (**d**) Upper shoot portion was cut and transferred; (**e**) Within 5 days from the base of shoot, root started emerging; (**f**) Measured length of root after 10 days of

proliferation of hairy roots was better on B5 liquid medium. Sen et al. (1993) cultured roots of *Coleus forskohlii* on different media under different cultural regimes. The roots grew well in the presence of all dosages of 4-Cl-IAA and 5,6-Cl_2-IAA with 25% MS and 1% sucrose and produced more forskolin as compared with IAA or IBA. The highest amount of forskolin (0.09%) was raised at 50 micrograms/l 4-Cl-IAA after 60 days of culture.

7.2.2 Shoot Cultures

For obtaining plants with higher forskolin content, Mander-Henger (1988) obtained callus induction on B5 medium and treated these calli with different doses of mutagenic UV. From these UV-treated calli, he could regenerate plants that were hardened and transferred to the soil. These regenerated plants showed variation in leaf morphology and arrangement. Compared to the untreated control, some regenerated plants showed a 1.5–2 fold increase in forskolin content.

In vitro clonal multiplication of *C. forskohlii* on MS medium fortified with Kn (2 mg/l) and IAA 1 mg/l) using nodal explants was investigated by Sharma et al. (1991). Shoots multiplied at a rate of 12-fold every 6 weeks.

Sen et al. (1992) investigated the production of forskolin in shoot tips, callus, and root tip cultures. Cultures were established on MS and White's basal media with and without PGRs. From shoot tip explants of aseptically germinated seedlings callus and multiple shoots were obtained. BAP at 0.5–2.5 mg/l promoted shoot induction but forskolin production was not affected. Callus was generated on MS medium and then transferred to White's medium with 1 mg/l BAP and 1 mg/l NAA where they became friable, whitish, and rhizogenic after 2 months. The PGR composition of the medium plays an important role in the production of secondary metabolites. DDF, another component in the root tubers of *C. forskohlii,* is pharmacologically inactive. DDF, also found in cell culture, can be transformed into active forskolin by introduction of the essential hydroxyl groups 1α and 9α by the fungus *Scophlariopsis* and chemical acetylation (Ganguli 1986; Nadkarni et al. 1986; Inamdar et al. 1989).

Tripathi et al. (1995) could produce of forskolin from root callus of *C. forskohlii.* Phytohormones, glycine, casein hydrolysate, and sucrose content in the medium influenced growth and production of forskolin. Biomass production was highest with 4 ppm NAA.

Mukherjee et al. (1996) made an attempt to determine the potential of tumorous cultures for the production of forskolin. They initiated transformed cultures of *C. forskohlii* using *A. tumefaciens.* They could obtain tumor tissue and shooty teratomas in in vitro cultures. Forskolin was detected in tumorous callus (0.002%),

---

Fig. 7.2 (continued) inoculation of cut shoot; (**g**) Transferred shoot started producing root in huge amount in MS basal medium; (**h**) Amount of root started increasing day by day; (**i**) Shoot portion also started producing root; and (**j**) Established hairy root culture in B5 medium (Source: Pandey et al. 2014) rhizogenic callus (0.11%), and root cultures (0.14%) but not in shooty teratomas. Mukherjee et al. (2003) reported an increase in the yield of forskolin in transformed roots, rhizogenic calli, and suspension cultures of *C. forskohlii* under the influence of IAA, IBA, NAA, 2,4-D, KN, BAP, GA_3, and auxin conjugates IAA-ala, IAA-gly, IAA-phe, and IAA-asp.

7.3 Unorganized Cultures

7.3.1 Callus Cultures

Swaroopa et al. (2016) investigated the response of different explants for callus induction. To obtain friable callus, cultures have been initiated from explants such as petiole, leaf, root, internode, and node of in vitro germinated plants. These explants were placed on MS, B5, and LS medium supplemented with 2 mg/l 2, 4-D to study the effect of media and explants on callus proliferation.

7.3.1.1 Effect of Sucrose on Callus Induction and Forskolin Content

The effect of various concentrations of sucrose on biomass yield and forskolin content was tested by Swaroopa et al. (2016). Sucrose not only influenced the nature of the callus and biomass production but also affected the forskolin content. Of the concentrations tested sucrose at 3% and 4% resulted in higher biomass accumulation, i.e., 86 gm and 89 gm (Fresh wt) within 15 days of initiation. However, maximum forskolin content was recorded in callus proliferated on medium supplemented with 4% sucrose. Swaroopa et al. (2016) reported that higher level of sucrose resulted in rhizogenic callus and increase in forskolin content.

7.3.1.2 Effect of Plant Growth Regulators on Callus Induction and Growth

Malathy and Pai (1999) obtained callus by culturing leaf, stem, and root on MS medium supplemented with auxins NAA, 2,4-D, IBA, and IAA. Forskolin content on stem callus was higher than on leaf callus.

Swaroopa et al. (2016) reported that highest percentage of callus induction was obtained on MS medium individually amended plant growth regulators such as 2, 4-D at 2 mg/l followed by NAA and IAA. Rhizogenesis on callus was observed on NAA and IAA supplemented medium. The cytokinins, BAP, and Kn at various concentrations and combinations were also tested by Swaroopa et al. (2016).

For callus induction, different combinations of growth regulators at different concentrations were also tested. Medium containing 3 mg/l NAA + 1 mg/l BAP resulted in highest percentage of callus induction (60.7 ± 5.8) and biomass production (89.5 ± 1.0). The combination in the medium resulted in high percentage of forskolin (184 ± 5.0 mg/kg DCW), but less biomass compared to 2 mg/l 2, 4-D (Swaroopa et al. 2016).

7.3.1.3 Growth and Forskolin Profile of Callus Cultures and Suspension Cultures

A time-course study of growth of callus cultures was conducted by Swaroopa et al. (2016) for 30 days on MS medium supplemented with 2, 4-D (2 mg/l) and 3% sucrose showed a sigmoidal curve. The fresh biomass was high in callus cultured on a solid medium. The growth cycle showed a lag phase of 4 days followed by an exponential phase of growth which resulted in increased growth till the 22nd day (39.73 gm FW) of culture and thereafter it reached the stationary phase (Swaroopa et al. 2016).

7.3.2 Suspension Cultures

Mersinger et al. (1988) reported forskolin production in untransformed suspension cultures of *C. forskohlii*. They reported that certain stages of differentiation in the direction of root formation are required for forskolin production. For initiation of callus cultures high producing plants of *C. forskohlii* were cultured on B5 medium fortified with 0.5 mg/l 2,4-D and 0.2 mg/l Kn under continuous light. To establish suspension cultures, the cultures were then transferred to B5 medium supplemented with different concentrations and combinations of 2,4-D, Kn, casein hydrolysate, IBA, or BAP. Forskolin production was observed in the cultures maintained on hormone-free medium or those supplemented with a combination of IBA and Kn or BAP. Maximum forskolin production (0.03%) was observed in medium containing 0.4 mg/l BAP alone followed by 0.025% in medium containing combination of IBA (0.4 mg/l) and Kn (0.2 mg/l). The suspension cultures lost their capacity to produce forskolin after 3–4 years and new cultures had to be established. With respect to forskolin production, the two cell strains cultured in continuous light or darkness (light and dark strains, respectively) showed different behavior. Forskolin production was higher in the light strain than in the dark strain. For the light strain, maximum forskolin production (0.15%) was observed in the second culture period (14 days) whereas the dark strain attained maximum production (0.05%) only in the fourth induction period.

Mukherjee (1998) and Mukherjee et al. (2000a) reported that the establishment of forskolin yielding transformed cell suspension cultures of *C. forskohlii* was controlled by different factors. Suspension cultures derived from gall calli which were obtained following infection with *Agrobacterium tumefaciens* (C58) were established in *C. forskohlii*. A fast-growing cell line (GSO-5/7) was found to accumulate 0.021% forskolin in 42 days.

7.3.2.1 Effect of Biotic Elicitors

Secondary metabolites are not produced in the required quantity in many cultures. Production of secondary metabolites can be enhanced by the treatment of the cultures with elicitors like chitosan, salicylic acid, methyl jasmonate, and heavy metals.

7.3.2.2 Elicitation of Forskolin Using Elicitors of Fungal Origin

Swaroopa et al. (2013b) studied the influence of *Aspergillus niger* in the production of forskolin in suspension cultures of *C. forskohlii*. The optimal time for the addition of *A. niger* for elicitation of forskolin accumulation was day 6 of the cultivation period. Concentration of elicitor varied between 5 and 20%. For assay of dry weight and forskolin content, cell cultures and cells were harvested 24 and 48 h after the addition of the elicitor. Forskolin content at 24 h after addition of elicitor (5% mycelial filtrate) was found to result in forskolin levels of 1178 mg/kg DW in the cells while in the control it was 220 mg/kg DW. Addition of 5% mycelial extract (ME) also resulted in forskolin content 315.3 mg/kg DW which was comparable to the control. Low levels of forskolin were observed with other levels of elicitors. Addition of ME and mycelia filtrate (MF) of *A. niger* to the cell suspension cultures of *C. forskohlii* resulted in a reduction in biomass accumulation.

The effect of *Fusarium oxysporum* as an elicitor was tested by Swaroopa et al. (2013b) by adding the ME and MF at concentrations of 5, 10, 5, and 20% to the suspension culture. They reported that forskolin levels increased for several hours after the addition of elicitor. A minimal time is required to induce the process of elevated secondary metabolite production. Some types of cellulose–glucans are better elicitors than chitin–glucan. In *C. forskohlii,* the oligosaccharides liberated from the cell wall of *Aspergillus* were the best elicitors, even though their identity is still unknown.

Different concentrations of ME and MF of *Rhizopus oryzae* were tested by Swaroopa et al. (2013b) on elicitation of forskolin content in suspension cultures of *C. forskohlii.* The ME elicitor (w/v) extracted from *R. oryzae* at a level of 15% exhibited 72% increase in the forskolin production after 24 h of elicitation over control cultures of *C. forskohlii.*

The effects of *Penicillium notatum* as elicitor at concentrations, 5, 10, 15, and 20% and exposure times on forskolin production were studied by Swaroopa et al. (2013b). A higher concentration of elicitor filtrate (15%) responded positively in terms of forskolin accumulation which was increased by threefold after 48 h of incubation.

7.3.2.3 Effect of Bacterial Elicitors

The effect on forskolin production in suspension cultures of *C. forskohlii* with different concentrations of cell and filtrates of *Staphylococcus aureus* were tested by Swaroopa et al. (2013a). As the concentration of both cell extract and filtrate of *S. aureus* increased there was a steady increase in forskolin levels from 1.4- to 3.5-fold in cell suspension cultures after 24 h. However, there were negative effects of all the treatments after 48 h of exposure.

Swaroopa et al. (2013a) tested the influence of different doses of *Bacillus subtilis* on forskolin production. When cultures were exposed to a 20% concentration of elicitor for 24 h, highest forskolin yield, above fourfold was achieved. When 15% CF was used there was a 3.3-fold enhancement of forskolin. *B. subtilis* at high concentration increased forskolin content, however, there was a decrease in biomass yield.

The effect of different concentrations of cells and filtrates of *Pseudomonas aeruginosa* was tested by Swaroopa et al. (2013a) on elicitation of forskolin production in suspension cultures of *C. forskohlii*. The treatment with 5% of *P. aeruginosa* CE resulted in a higher level of total forskolin (635 gm/kg DCW) compared to the control (174 ± mg/kg DCW) after 24-h administration of the elicitor. There was an increase in forskolin production by 173 and 118% in comparison to the control after 24-h administration at concentrations 10 and 15% of CE, respectively. There was no reduction in biomass production at lower concentrations but at higher concentrations there was a slight decrease in biomass yield.

On day 6 of the cultivation period, different concentrations (5 and 20%) of *E. coli* elicitor were added to suspension cultures of *C. forskohlii* for elicitation of forskolin accumulation (Swaroopa et al. 2013a). The addition of 15% CF was found to result in higher forskolin levels (547.2 ± 2.3 mg/kg DW) in the cells which were comparable to the control (174.0 ± 1.0 mg/kg DW) at 24 h after addition of elicitor. CF at 10% also resulted in higher forskolin content (703.7 ± 7.5 mg/kg DW), which was comparable to the control. At 5 and 20%, the forskolin levels were low compared to 10 and 15% CF of *E. coli*.

The effect of *Proteus aureus* was tested by Swaroopa et al. (2013a) by adding the CE and CF at concentrations of 5, 10, 5, and 20% and comparing the results with a control without elicitor. Twenty-four hours after elicitation at 10% of CE forskolin contents was 610 mg/kg of the cell dry weight in comparison to the control 174 mg/kg of the cell dry weight.

7.3.3 Effect of Microbial Derived Elicitors

The effect of casein hydrolysate on the cell suspension cultures of *C. forskohlii* was studied by Swaroopa (2008). Different concentrations (0.2, 0.4, and 0.6 gm) of casein hydrolysate were tested to optimize the concentration for elicitation of forskolin. Of all the concentrations, 0.4 gm of casein hydrolysate has shown a 285% increase in forskolin accumulation compared to control cultures (100%). However, there was a significant difference in response with respect to different concentrations of casein hydrolysate with forskolin content at 24 h and 48 h. There was no literature regarding the casein hydrolysate as an elicitor in elicitation of secondary metabolites.

Swaroopa (2008) reported that addition of 1% (w/v) yeast extract to the *C. forskohlii* cells enhanced the forskolin content by 398% and at 2% (w/v) level there is an increase by 313% in comparison to control after 48-h elicitation. Above this level, decreased forskolin content was observed in comparison to control. And again at 5% and 6%, the forskolin content increased by 47% and 65%, respectively. Variation in forskolin production is often observed and the mechanism is not yet clear. There was a reduction in biomass production at higher concentrations.

The effects of different concentrations (2, 4, 6, and 8 gm) of sodium alginate were tested by Swaroopa (2008) on the forskolin content and biomass yields. There was a 1.4-fold increase in forskolin content in comparison to control cultures at 2 gm and cell growth was inhibited by using sodium alginate as an elicitor. This may be due to the effect of alginate, which modifies the medium viscosity leading to poor oxygen transfer.

7.3.4 Effect of Abiotic Elicitors

Metal ions that are already present in the standard MS medium in trace amounts were chosen by Swaroopa et al. (2015) for testing elicitation capability. The concentration, incubation time, and the extent of elicitation varied with metal ions tested. Calcium at 100 mM concentration at 24 h after addition of elicitor enhanced forskolin levels to 180% over the control. The other concentrations had shown negative effects on forskolin production.

Different concentrations of $CuSO_4$ (0.25–1.00 mM) and $ZnSO_4$ (25–100 mM) were tested by Swaroopa et al. (2015). Cu and Zn at concentrations of 0.50 mM after 48-h incubation and 50 mM after 24-h incubation enhanced forskolin production by 142% and 59%, respectively. There was no significant effect of remaining concentrations tested on forskolin production. Suspension cultures treated with different concentrations of $CuSO_4$ and $ZnSO_4$ showed significant difference at 5% level, with forskolin content at 24 h and 48 h, respectively. Finally, neither of the microelements tested had effect on biomass yields. It was not observed that increasing trace metal concentration induced the excretion of forskolin of *C. forskohlii*. This effect might be attractive since forskolin is vacuolar metabolite, a situation making difficult their excretion.

7.3.5 Effect of Plant-Derived Elicitors

The four concentrations of salicylic acid (25, 50, 75, and 100 mM) in the production of forskolin by means of suspension cultures were examined after 24- and 48-h administration of the elicitor (Swaroopa et al. 2015). In cultures treated with 0.5 mM of salicylic acid, there was a 2.4-fold enhancement in the forskolin content in the cells compared to the control after 24-h administration of the elicitor. Suspension cultures treated with different concentrations of salicylic acid showed significant differences at 5% level, with forskolin content at 24 h and 48 h.

The effects of different concentrations of acetylsalicylic acid (0.5, 1.0, 1.5, and 2.0 mM) were tested by Swaroopa et al. (2015) on the forskolin content and biomass yields. There was twofold increase in forskolin content in comparison to control cultures at 0.5 mM after 48-h elicitation and cell growth was inhibited at higher concentrations. Statistical F-Test at 5% level indicates a significant difference in response with respect to different concentrations of acetylsalicylic acid at 24 h and 48 h.

When the cells were treated with different concentrations of methyl jasmonate (25–100 mM) resulted in a higher level of total forskolin (650 ± mg/kg DCW) compared to the control (148 ± mg/kg DCW) after 24-h (Swaroopa et al. 2015).

Forskolin production after 24-h administration at concentrations 50 and 75 mM, increased by 170% and 68% in comparison to the control, respectively. Biomass production at lower concentrations was not observed. When the concentration is increased above this resulted in a decrease in biomass and forskolin production than control. Suspension cultures treated with different concentrations of methyl jasmonate showed significant differences at 5% level, with forskolin content at 24 h and 48 h.

Various levels of guar gum were added to the culture media and the effect on the suspension cultures of *C. forskohlii* was studied by Swaroopa (2008). The maximum forskolin production of 1214 gm/kg was observed in cells treated with 50 μg guar gum. There was ninefold enhancement in forskolin content in comparison to control cultures after 24 h of elicitation. When the concentration was raised to 100 μg there were 6.4- and 4.2-fold at 150 μg of guar gum. This enhancement is observed after 24-h exposure to the elicitor. The other higher levels of guar gum were found to lower forskolin content. There was no effect seen on biomass yield when treated with guar gum. Guar gum is the best elicitor on yield enhancement of forskolin in cell suspension cultures of *C. forskohlii*. The data subjected for statistical analysis indicate a significant difference at 5% level in cultures subjected to various concentrations of guar gum in terms of forskolin content at 24 h and 48 h.

References

Asada Y, Li W, Terada T, Xinzhu LQ, Yoshikawa T, Hamaguchi S, Namekata I, Tanaka H, Koike K (2012) Labdane-type diterpenoids from hairy root cultures of *Coleus forskohlii*, possible intermediates in the biosynthesis of forskolin. Phytochemistry 79:141–146

Balandrin MJ, Klocke JA (1988) Medicinal, aromatic and industrial materials from plants. In: Bajaj YPS (ed) Biotechnology in agriculture and forestry, Medicinal and aromatic plant, vol 4. Springer-Verlag, Berlin, Heidelberg, pp 1–36

Ganguli BN (1986) Microbial transformation of forskolin derivatives. In: Rupp RH, de Souza NJ, Dohadwalla AN (eds) Forskolin: its chemical, biological and medicinal potential. Hoechst India, Bombay, pp 31–38

Inamdar PK, Khandelwal Y, Garkhedkar M, Rupp RH, de Souza NJ (1989) Identification of microbial transformation products of 1,9 dideoxyforskolin and 7-deacetyl-1, 9-dideoxyforskolin. Planta Med 55:386–387

Krombholz R, Mersinger R, Kreis W, Reinhard E (1992) Production of forskolin by axenic *Coleus forskohlii* roots cultivated in shake flasks and 20-l glass jar bioreactors. Planta Med 58:328–333

Krombholz R, Mersinger R, Reinhard E (1990) Root cultures of *Coleus forskohlii*: a system for continuous production of forskolin. Planta Med 56(6):629–630

Kukreja AK, Garg S (2007) *Agrobacterium rhizogenes* mediated hairy roots in *Coleus forskohlii* – a natural source of labdane diterpenoid (forskolin). In: Trivedi PC (ed) Biotechnology – current perspectives and potential applications. Avishkar Publications, Distributors, Jaipur, pp 298–346

Li W, Koike K, Asada Y, Yoshikawa T, Nikaido T (2005) Rosmarinic acid production by *Coleus forskohlii* hairy root cultures. Plant Cell Tissue Organ Cult 80:151–155

Malathy S, Pai JS (1999) Monitoring of forskolin production from roots and callus by HPTLC in *Coleus forskohlii* Briq. J Spices Arom Crop 8:153–157

Mander-Henger A (1988) Somaclonal variation bei *Coleus forskohlii*. Doctoral thesis. University of Tubingen

Mersinger R, Dornauer H, Reinhard E (1988) Formation of forskolin by suspension cultures of *Coleus forskohlii*. Planta Med 54:200–204

Mukherjee S (1998) *Agrobacterium* mediated transformation in medicinal plants and studies on relationship between organisation and secondary metabolite production in vitro. Ph.D. Thesis. Calcutta University

Mukherjee S, Ghosh B, Jha S (1996) Forskolin synthesis in in vitro cultures of *Coleus forskohlii* Briq. transformed with *Agrobacterium tumefaciens*. Plant Cell Rep 15:691–694

Mukherjee S, Ghosh B, Jha S (2000a) Establishment of forskolin yielding transformed cell suspension cultures of *Coleus forskohlii* as controlled by different factors. J Biotechnol 76 (1):73–81

Mukherjee S, Ghosh B, Jha S (2000b) Enhanced forskolin production in genetically transformed cultures of *Coleus forskohlii* by casein hydrolysate and studies on growth and organisation. Biotechnol Lett 22(2):133–136

Mukherjee S, Ghosh B, Jha S (2002) Genetic transformation of *Coleus forskohlii* for forskolin production. In: Nandi SK et al (eds) Role of plant tissue culture in biodiversity, conservation and economic development. Gyanodaya Prakashan, Nainital, pp 277–286

Mukherjee S, Ghosh B, Jha S (2003) Higher production of forskolin in genetically transformed cultures of *Coleus forskohlii* Briq. induced by growth regulators. J Plant Biochem Biotech 12:81–85

Nadkarni SR, Akut PM, Ganguli BN, Khandelwal Y, de Souza NJ, Rupp RH, Fehlhaber H (1986) Microbial transformation of 1,9-dideoxyforskolin to forskolin. Tetrahedron Lett 27:5265–5268

Nagpal A, Singh B, Sharma S, Rani G, Virk GS (2008) *Coleus* spp.: micropropagation and in vitro production of secondary metabolites. Med Arom Plant Sci Biotechnol 2:1–17

Pandey R, Krishnswamy V, Kumaravadivel N, Rajamani K (2014) Establishment of hairy root culture and production of secondary metabolites in Coleus (*Coleus forskohlii*). J Med Plant Res 8:58–62

Petersen M (1994) *Coleus* spp.: in vitro culture and the production of forskolin and rosmarinic acid. Med Aroma Plants 4:69–92

Reddy CS, Praveena C, Veeresham C (2012) Strategies to improve the production of forskolin from hairy root cultures of *Coleus forskohlii* Briq. Int J Pharm Sci Nanotechnol 5:1720–1726

Sasaki K, Udagawa A, Ishimaru H, Hayashi T, Alfermann AW, Nakanishi F, Shimomura K (1998) High forskolin production in hairy roots of *Coleus forskohlii*. Plant Cell Rep 17:457–459

Sen J, Sharma AK, Mahato SB (1993) Forskolin production in untransformed root culture of *Coleus forskohlii*. Phytochemistry 34(5):1309–1312

Sen J, Sharma AK, Sahu NP, Mahato SB (1992) Production of forskolin in in vitro cultures of *Coleus forskohlii*. Planta Med 58:324–327

Sharma N, Chandel KPS, Srivastava VK (1991) *In vitro* propagation of *Coleus forskholii* Briq. A threatened medicinal plant. Plant Cell Rep 10:67–70

Swaroopa G (2008) Elicitation of forskolin in suspension cultures of *Coleus forskohlii* (Willd.). Briq. PhD thesis. Acharya Nagarjuna University, Guntur

Swaroopa G, Anuradha M, Pullaiah T (2013a) Elicitation of forskolin in suspension cultures of *Coleus forskohlii* (Willd.) Briq. using bacterial elicitors. J Indian Bot Soc 92:97–100

Swaroopa G, Anuradha M, Pullaiah T (2013b) Elicitation of forskolin in suspension cultures of *Coleus forskohlii* (Willd.) Briq. using elicitors of fungal origin. Curr Trends Biotech Pharm 7 (3):755–762

Swaroopa G, Anuradha M, Pullaiah T (2015) Elicitation of forskolin in suspension cultures of *Coleus forskohlii* (Willd.) Briq. using abiotic elicitors. J Trop Med Plants 15:81–86

Swaroopa G, Anuradha M, Pullaiah T (2016) Influence of sucrose and plant growth regulators on growth and forskolin production in callus cultures of *Coleus forskohlii*. Curr Trends Biotech Pharm 10(1):45–54

Tripathi CKM, Basu SK, Jain S, Tandon JS (1995) Production of coleonol (forskolin) by root callus cells of plant *Coleus forskohlii*. Biotechnol Lett 17:423–426

Uma Maheswari R, Selvamurugan C, Jayabarath J, Lakshmi Prabha A (2011) Hairy root culture of an important medicinal plant: *Coleus forskohlii*. Int J Agric Sci 3(2):82–89

Assessment of *Coleus forskohlii* Genetic Diversity Using Molecular Markers

8

Abstract

Genetic variation is indispensable for the long-standing survivability of species, and also it plays an important role in conservation. Molecular markers provide assistance in the development of more dependable tools for selecting and conservative approaches for many endemic plant species. For improving crops and their yields, molecular diversity evaluation and detection of superior genotypes are very crucial. Previously, simple morphological markers were employed to evaluate the molecular diversity and to identify superior genotypes. However, morphological characters tend to vary with the surrounding ecological circumstances. Thus, morphological markers have greater limitations as compared to molecular markers that depend on the DNA sequences. *C. forskohlii* (syn. *Plectranthus barbatus*) is one of the most prevalent medicinal plant, and is conventionally used for treating various conditions, such as asthma, bronchitis, hay fever, and stomach disorders. Forskolin is the chief bioactive metabolite occurring exclusively in *C. forskohlii* roots, and it is utilized in the management of asthma, glaucoma, cardiovascular disorders, and gastrointestinal problems. In this chapter, the assessment of *C. forskohlii* genetic diversity using diverse molecular markers is presented in detail.

Keywords

RAPD · AFLP · ISSR · DNA fragments · genetic diversity · UPGMA · Molecular markers

8.1 Introduction

Plants as sessile creatures yield plentiful chemocompounds that are together renowned as specific metabolites (Swamy and Akhtar 2019; Swamy 2020). Though these plant-derived compounds are not essential for plant's growth and development,

T. Pullaiah, *Forskolin*, https://doi.org/10.1007/978-981-19-6521-0_8

they are mainly secreted against biotic and abiotic stress conditions. These metabolites are particularly important as defense molecules against attacks by biotrophic pathogens and herbivores (Akhtar and Swamy 2018; Swamy and Akhtar 2019; Swamy et al. 2019; Karthikeyan et al. 2020). These compounds belong to different chemical classes, including phenolics, alkaloids, terpenes, steroids, and vanilloids. Extraordinarily, these metabolites exhibit numerous therapeutic properties, and hence have become one of the sought-after molecules of choice for the treatment of various ailments in humans (Swamy 2020). So far, many plant species have been recognized and explored for their medicinal uses. Among them, a few species belonging to the genus, *Coleus* (family; Lamiaceae) includes several herbal plant species that are extensively found in several parts of the Asian countries, particularly in India, and have shown to possess a significant commercial importance with respect to their tonic and nutraceutical values (Kavitha et al. 2010; Mitra et al. 2020). *Coleus forskohlii* (syn. *Plectranthus barbatus*) is one of the most prevalent medicinal plant, and is conventionally used for treating various conditions, such as asthma, bronchitis, hay fever, and stomach disorders. Forskolin is the chief bioactive metabolite occurring exclusively in *C. forskohlii* roots, and it is utilized in the management of asthma, glaucoma, cardiovascular disorders, and gastrointestinal problems. The plant's root extract has the potential to be used as an adjunct treatment for the management of obesity. Forskolin has been shown to improve individual wellness by supporting lean body mass and promoting healthy metabolic functions (Kamohara and Noparatanawong 2013). Further, antimicrobial properties are exhibited by the essential oil obtained from the plant's tubers (Alasbahi and Melzig 2010; Mitra et al. 2020; Singh et al. 2011; Jagtap et al. 2011). A wide-ranging bioactive metabolites are known to occur in this plant. Some of the major ones include barbatusin, methyl quinine, coleonol, coleon, coleosol, phenols, glycosides, and terpenoids (Kavitha et al. 2010). Thus, this plant is used in numerous fields, including pharmaceutical firms to produce several medicines. In the food industry, it is being used primarily as a condiment. The plant's tubers are used in the preparation of edible pickles (Wagh et al. 2014; Kavitha et al. 2010; Mitra et al. 2020).

Genetic variation is indispensable for long-standing survivability of species, and also it plays an important role in conservation. The occurrence and evolution of plant species depend on their genomic diversity. This helps in species adaptation to varying circumstances and a number of abiotic and biotic stresses (Indu et al. 2019). Genetic variability is one of the criteria for selection or crop improvement programs (Hohenlohe et al. 2021; Swarup et al. 2021). Hence, it is essential to identify and obtain the data related to variation occurring between and within the population. Also, for proficient protection and managing, the species' genetic configuration in diverse topographical localities needs to be evaluated (Govarthanan et al. 2011; Crossa et al. 2021). The conservation of biodiversity is the crucial concept in maintaining the wild gene pool. Plant conservation methods basically depend on general principles, viz., leading to a standardized system for the assessment of the disappearance or under-threat plant species (Morais et al. 2014; Penas et al. 2016; Indu et al. 2019). More recently, efforts have been put into improving in situ and ex situ conservation of vital medicinal plant species using different strategies. These conservation tactics often require complete knowledge of population dynamics, and report on the structural genetics and diversity intensities among species (Heywood 2014; Perez-Collazos et al. 2008; Indu et al. 2019; Volis and Blecher 2010). The occurrence of genetic diversity can be assessed by employing different molecular markers that are highly specific, and can effectively distinguish plants from their phenological traits. Understanding the genetic diversity levels can be very useful in medicinal plant breeding programs. Furthermore, they provide assistance in the development of more dependable tools for selecting and conservative approaches for many endemic plant species (Indu et al. 2019).

The genomic diversity levels can be normally ascertained by making use of various molecular markers that are highly specific and distinguish plants from the phenological disparities. Additionally, these markers assist in the development of consistent and more effective approaches to selecting and conserving many rare plant species (Jaisankar et al. 2017). Some of the combined approaches, including genetic markers, geographical information systems, and bioinformatics are helpful for developing a few appropriate techniques to survey, collect, and assessment of plant genomic diversities (Indu et al. 2019). The major molecular markers include random amplified polymorphic DNAs (RAPDs), restriction fragment length polymorphisms (RFLPs), amplified fragment length polymorphism (AFLP), inter-simple sequence repeat (ISSR), single-stranded conformation polymorphism (SSCP), and single sequence repeats (SSRs). They are widely used to establish species or identification of varieties, identification of genetic diversity and for gene mapping and also to know about phylogenetic relationships (Jaisankar et al. 2017). For detecting genetic variations among individuals, molecular markers allow absolute discovery of the gene of importance, and assists linkage study by assisting in genetic map assembly (Lidder and Sonnino 2012). In this chapter, the assessment of *C. forskohlii* genetic diversity using diverse molecular markers is presented in detail.

8.2 Molecular Markers

In genetics, molecular markers, also known as genetic markers are DNA fragments that are linked with definite sites within the genome. In molecular biology, these DNA-based markers are utilized for identifying particular DNA sequences in a group of unidentified DNA. Additions, deletions, point mutations, duplications, and translocations are the basis for DNA polymorphisms. However, these polymorphisms may not essentially affect the gene's activities (Indu et al. 2019). The best DNA marker should be codominant, uniformly distributed in the genome, have the capability to identify increased levels of polymorphisms, and highly reproducible (Mondini et al. 2009; Madhumati 2014; Nadeem et al. 2018). These markers are categorized based on: 1. modes of gene actions (dominant or codominant markers); 2. methods of identification (PCR (Polymerase Chain Reaction)-based markers or hybridization-centered DNA markers); and 3. modes of transmissions (parental organelle inheritance, maternal nuclear inheritance) (Semagn et al. 2006; Nadeem et al. 2018). Different types of DNA molecular markers have been developed and successfully applied in genetics and breeding activities in various agricultural crops.

For improving crops and their yields, molecular diversity evaluation and detection of superior genotypes are very crucial. Previously, simple morphological markers were employed to evaluate molecular diversity and to identify superior genotypes. However, morphological characters tend to vary with the surrounding ecological circumstances. Thus, morphological markers have greater limitations as compared to molecular markers that depend on the DNA sequences (Kavitha et al. 2009; Swamy and Anuradha 2011). In the following section, molecular marker studies related to only *C. forskohlii* are discussed.

8.3 Molecular Marker Studies in *Coleus forskohlii*

For the first time, RAPD technique was used to evaluate the genetic diversity in *C. forskohlii* in 2009 (Kavitha et al. 2009). Thirty-seven genotypes of *C. forskohlii,* sampled from various regions of Karnataka and Tamil Nadu were used in their diversity study using 25 RAPD primers. The results showed the occurrence of 117 DNA bands, and among them 60 bands were found to be polymorphic (51.28%), yielding an average of 3.75 bands per primer. The total number of bands for primers varied between 1 and 7. The obtained data of RAPD analysis was subjected to cluster analysis on Jaccard's similarity coefficient matrices. A dendrogram of 37 genotypes of *C. forskohlii* was generated, revealing 2 major clusters. Overall, their observations directed that RAPD may perhaps be employed for molecular diversity studies in *C. forskohlii* involving higher number of primers as it is consistent, easy, quick, and inexpensive.

A study demonstrated the impact of molecular and ecological factors in producing secondary metabolites in *C. forskohlii* (Revadigar et al. 2008). They used fresh and dried roots of *C. forskohlii* collected from seven regions in India. RAPD analysis data of *C. forskohlii* accessions exhibited diverse levels of genetic variation. A total of 193 DNA fragments were amplified from the 20 primers used in their investigation. Among these, 182 fragments were polymorphic (83%) in at least 1 of the 7 accessions. The polymorphism levels varied with primers for these accessions. The primers, namely BGA1 and BGA10 yielded maximum numbers of amplified fragments, i.e., 26 and 34, respectively. The primers BGA6 and BGA7 resulted in only three amplified products. Overall, the amplified products number from each accession differed considerably for all the tested primers. Five primers, namely BGD7, BGD6, BGDA9, BGD9, and BGDA8 failed to yield any amplified products. The similarity matrix attained through Jaccard's approach exposed a coefficient of similarity values that ranged between 0.23 and 0.58 with a mean value of 0.41. UPGMA analysis revealed two main clusters encompassing Hyderabad, Bangalore, and Arabhavi in the first cluster and Baroda, Salem, and Akot in the second cluster. Whereas, Dharwad accession was found to align separately. These results suggested the existence of higher genetic variability amongst these accessions. Among the seven accessions, Hyderabad and Bangalore origins exhibited more similarity (0.58) as compared to others. While, Dharwad showed the least similarity (0.23) with others, suggesting that it is genetically different from other accessions.

A study aimed to evaluate the genetic differences among *C. aromaticus*, *C. amboinicus,* and *C. forskohlii* using the RAPD approach. Also, it determined the levels of molecular similarity amongst them (Govarthanan et al. 2011). The highest number of DNA bands were recorded for *C. amboinicus* when OPW 6 primer was used. However, no amplification was observed for *C. aromaticus.* In *C. forskohlii,* 13 bands were amplified. Overall, variations among *Coleus* species were detected from OPW (6–10) series primers, however, OPU (15–19) series primers failed to show differences. Overall, the results obtained in this study showed that it is a prerequisite to include individuals from more diverse populations in order to conserve their biodiversity for the future.

A concurrent, metabolic, and molecular diversity study was performed by Ahmad et al. (2013). They employed high-performance thin-layer chromatography (HPTLC) fingerprinting and RAPD markers to evaluate chemo- and genetic diversity among ten different *C. forskohlii* genotypes samples. Twenty RAPD primers were used for evaluating genetic variations. They used pre-coated aluminum back thin-layer chromatography sheets as stationary phase and 3 different mobile phases for chromatographic analysis. A genetic correlation matrix was created by means of chromatographic and molecular fingerprint information on the basis of Jaccard's coefficient and additional acquaintance to develop a dendrogram based on the unweighted pair group method with arithmetic mean (UPGMA). According to the results, the similarity coefficient for genetic nature was found to be between 0.362 and 0.783, while for metabolic profile, it was between 0.683 and 0.939. This indicated the occurrence of molecular diversity in the samples of *C. forskohlii,* however, with a less significant chemo-diversity among the same samples.

AFLP, RAPD, and ISSR markers were employed for the assessment of morphological and genetic diversity in 18 different *C. forskohlii* genotypes were sampled from various locations in central India (Tripathi et al. 2013). About 80, 101, and 483 fragments were produced from 10 ISSR, 11 RAPD, and 8 AFLP primers, correspondingly. Among the 3 markers employed in their study, ISSR and RAPD showed polymorphism of about 68 and 61%, respectively. While, 8 AFLP primers resulted in more than 70% of polymorphism. UPGMA cluster analysis approach assembled genotypes in two groups after considering all DNA marker methods independently, and after pooled examination. Overall, these results showed the usefulness of these morphological and molecular markers in observing variations in *C. forskohlii* genotypes.

In another study, 12 *C. forskohlii* germplasms from various populations were evaluated using ISSR marker for genetic fingerprinting, and quantification of forskolin was performed (Srivastava et al. 2017). Two selected germplasms, namely NBC-16 (0.641%) and NBC-24 (0.728%) accumulated the maximum forskolin content with greater molecular variability of 92%. UPGMA hierarchical clustering arrays exposed strong molecular consortium among the individuals agreeing to their topographical collections. Mantel testing exposed a positive association between genetic and biochemical fingerprints, reflecting the ISSR marker's viability to analyze genome data interrelated to the biosynthesis of forskolin from individuals from different phytogeography. An affirmative relationship between forskolin content and altitude rise was also denoted using the Pearson correlation coefficient. Nevertheless, the relationship of both molecular and biochemical fingerprint information with the phytogeographic distance matrix was actually negative, and it predestined that distance may be a predictor of inhabitants' diversity. This study signified the usefulness of molecular and metabolic fingerprints in identifying elite accessions and affords evidence to industries to commercially exploit *Coleus* species, comprising their locality-specific profitable farming.

Lately, researchers have examined the influence of kinetin treatments on the biochemical contents of five *Coleus* cultivars. Also, they used ISSR markers to evaluate their molecular diversity (Shoaib et al. 2020). The results showed that *Coleus* cultivars had ample quantities of chlorophyll pigments (CP), total carotenoids (TC), and phytochemicals, such as total phenolics (TP), total carbohydrates (TCs), total flavonoids (TF), and total tannins (TT). Among all the cultivars, C5 (Finger paint) cultivar possessed the maximum amounts of CP, TC, TCs, TP, and TT. However, the highest TF contents were recorded in C2 (Beckwhites cultivar) (11.814 mg/g) and C5 cultivars (11.648 mg/g). ISSR marker examination was executed on the *Coleus* cultivars using five primer pairs. The results showed the occurrence of 18 monomorphic bands and 21 polymorphic bands with polymorphism (53.8%). Overall, 10 distinctive bands were recognized from the resulting ISSR profile. These outcomes advocated that the influence of kinetin usages encourages secondary metabolite secretion. Further, the ISSR marker can be very useful in likely selection, ascertaining, distinguishing, and handling of *Coleus* cultivars.

8.4 Conclusions

Molecular markers have significantly helped in identifying elite accessions of *C. forskohlii* and give confirmations to manufacturing firms to commercially exploit *Coleus* species, considering their area-specific profitable farming. To date, for the molecular characterization and identification of *C. forskohlii*, RAPD marker is the most commonly used molecular marker. Unfortunately, DNA marker studies are majorly conducted on the *Coleus* population of Indian origin. Hence, similar attention should also be given to *Coleus* population occurring in other parts of the world. For a clear understanding on the polymorphism levels and genetic status of the *C. forskohlii* inhabitants all over the world, an increased number of DNA-based genetic (both dominant and codominant) marker studies should be encouraged. These genetic markers are indisputably treasured instruments for addressing plant breeding issues and population genetics. Nevertheless, for reconstructing the phylogeny and taxonomy studies, DNA markers may occasionally mislead, and hence they should be utilized with care. Future investigations should be focused toward the exploration of other unused genetic markers and NGS (next-generation sequencing) platforms in *C. forskohlii*. This will be useful in authenticating cultivars, conservation, and molecular diversity understanding and assisting breeding programs in *C. forskohlii*. Also, these available molecular markers should be used as a tool to overcome the problems of *C. forskohlii* adulteration.

References

Ahmad S, Tamboli E, Parveen R (2013) Determination of metabolic and genetic diversity in *Coleus forskohlii* Briq. using HPTLC and RAPD analysis. Planta Med 79(05):P103

Akhtar MS, Swamy MK (2018) Anticancer plants: properties and application. Springer

Alasbahi RH, Melzig MF (2010) *Plectranthus barbatus*: a review of phytochemistry, ethnobotanical uses and pharmacology–part 1. Planta Med 76(07):653–661

Crossa J, Fritsche-Neto R, Montesinos-Lopez OA, Costa-Neto G, Dreisigacker S, Montesinos-Lopez A, Bentley AR (2021) The modern plant breeding triangle: optimizing the use of genomics, phenomics, and enviromics data. Front Plant Sci 12:651480

Govarthanan M, Arunapriya S, Guruchandar A, Selvankumar T, Gnanasekaran N, Koildhasan M (2011) Genetic variability among *Coleus* sp studied by RAPD banding pattern analysis. Nat Prec:1. https://doi.org/10.1038/npre.2011.5631

Heywood VH (2014) An overview of in situ conservation of plant species in the Mediterranean. Flora Mediterranea 24:5–24

Hohenlohe PA, Funk WC, Rajora OP (2021) Population genomics for wildlife conservation and management. Mol Ecol 30(1):62–82

Indu B, Mohonty SK, Bhat S, Swamy MK, Anuradha M (2019) Genetic diversity and conservation of *Pterocarpus santalinus* L.f. through molecular approaches. In: Pullaiah T, Balasubramanya S, Anuradha M (eds) Red sanders: silviculture and conservation. Springer, pp 173–187

Jagtap M, Chandola H, Ravishankar B (2011) Clinical efficacy of *Coleus forskohlii* (Willd.) Briq. (Makandi) in hypertension of geriatric population. Ayu 32(1):59

Jaisankar I, Subramani T, Velmurugan A, Singh AK (2017) Genetic diversity revealed among ratstan genotyles from Andaman and Nicobar Islands based on RAPD and ISSR markers. Forest Res Eng Int J 1(2):45–50

Kamohara S, Noparatanawong S (2013) A *coleus forskohlii* extract improves body composition in healthy volunteers: an open-label trial. Personal Med Univ 2:25–27

Karthikeyan G, Swamy MK, Viknesh MR, Shurya R, Sudhakar N (2020) Bioactive Phytocompounds to fight against antimicrobial resistance. In: Plant-derived bioactives. Springer, pp 335–381

Kavitha C, Rajamani K, Vadivel E (2010) *Coleus forskohlii* a comprehensive review on morphology, phytochemistry and pharmacological aspects. J Med Plants Res 4(4):278–285

Kavitha C, Vadivel E, Sivasamy R, Rajamani K (2009) Analyzing variability in *Coleus forskohlii* Briq. using RAPD markers. J Horti Sci 4(2):143–147

Lidder P, Sonnino A (2012) Biotechnologies for the management of genetic resources for food and agriculture. Adv Genet 78:1–167

Madhumati B (2014) Potential and application of molecular markers techniques for plant genome analysis. Int J Pure App Biosci 2(1):169–188

Mitra M, Gantait S, Mandal N (2020) *Coleus forskohlii:* advancements and prospects of in vitro biotechnology. Appl Microbiol Biotechnol 104(6):2359–2371

Mondini L, Noorani A, Pagnotta MA (2009) Assessing plant genetic diversity by molecular tools. Diversity 1(1):19–35

Morais VA, Haddad D, Craessaerts K, De Bock P-J, Swerts J, Vilain S, Aerts L, Overbergh L, Grünewald A, Seibler P (2014) PINK1 loss-of-function mutations affect mitochondrial complex I activity via NdufA10 ubiquinone uncoupling. Science 344(6180):203–207

Nadeem MA, Nawaz MA, Shahid MQ, Doğan Y, Comertpay G, Yıldız M, Hatipoğlu R, Ahmad F, Alsaleh A, Labhane N (2018) DNA molecular markers in plant breeding: current status and recent advancements in genomic selection and genome editing. Biotechnol Biotechnol Equip 32(2):261–285

Penas J, Barrios S, Bobo-Pinilla J, Lorite J, Martínez-Ortega MM (2016) Designing conservation strategies to preserve the genetic diversity of *Astragalus edulis* Bunge, an endangered species from western Mediterranean region. PeerJ 4:e1474

Perez-Collazos E, Segarra-Moragues JG, Catalan P (2008) Two approaches for the selection of relevant genetic units for conservation in the narrow European endemic steppe plant *Boleum asperum* (Brassicaceae). Biol J Linn Soc 94(2):341–354

Revadigar V, Shashidhara S, Rajasekharan P, Pradeep N, Prakashkumar R, Murali B (2008) Variability in the chemical constituents in the roots of *Coleus forskohlii* from different geographical regions of India. Acta Hortic 765:245

Semagn K, Bjørnstad Å, Ndjiondjop M (2006) An overview of molecular marker methods for plants. Afr J Biotechnol 5(25):2540–2568

Shoaib RM, Ali SI, Metwally SA, Ibrahim MM, Aboud KA (2020) Phytochemical and molecular analyses of some *Coleus* cultivars cultivated in Egypt. Bull Natl Res Centre 44(1):1–8

Singh R, Gangwar SP, Singh D, Singh R, Pandey R, Kalra A (2011) Medicinal plant *Coleus forskohlii* Briq.: disease and management. Med Plants Int J Phytomed Relat Ind 3(1):1–7

Srivastava S, Misra A, Mishra P, Shukla P, Kumar M, Sundaresan V, Negi KS, Agrawal PK, Rawat AKS (2017) Molecular and chemotypic variability of forskolin in *Coleus forskohlii* Briq., a high value industrial crop collected from western Himalayas (India). RSC Adv 7(15):8843–8851

Swamy MK (2020) Plant-derived bioactives: production, properties and therapeutic applications. Springer Nature

Swamy MK, Akhtar MS (2019) Natural bio-active compounds. Springer

Swamy M, Anuradha M (2011) Analysis of genetic variability in patchouli cultivars (*Pogostemon cablin* Benth.) by using RAPD markers. Res Biotechnol 2(6):64–71

Swamy MK, Patra JK, Rudramurthy GR (2019) Medicinal plants: chemistry, pharmacology, and therapeutic applications. CRC Press

Swarup S, Cargill EJ, Crosby K, Flagel L, Kniskern J, Glenn KC (2021) Genetic diversity is indispensable for plant breeding to improve crops. Crop Sci 61(2):839–852

Tripathi N, Saini N, Tiwari S (2013) Morphological and molecular characterization of endangered medicinal plant species *Coleus forskohlii* collected from Central India. J Crop Sci Biotechnol 16(4):253–261

Volis S, Blecher M (2010) Quasi in situ: a bridge between ex situ and in situ conservation of plants. Biodivers Conserv 19(9):2441–2454

Wagh TM, Ptil S, Gangadharappa P, Prabhuling G, Shirol A, Naik R, Krishnamurthy G (2014) Performance of yield characters and suitability for pickling in different medicinal *Coleus* [*Plectranthus forskohlii* (Willd) Briq.] genotypes. Indian J Ecol 41(2):336–338

9 Endophytes for the Enhanced Growth of *Coleus forskohlii* and Enhanced Production of Forskolin

Abstract

Plant–fungi interaction renders enhanced production of plant secondary metabolites as most of these compounds are produced due to activation of defense-related biosynthetic pathways.

Fungal endophytes of *Coleus forskohlii* like *Piriformospora indica, Fusarium redolens*, *Phialemoniopsis cornearis*, and *Macrophomina pseudophaseolina* were used for in planta enhancement of forskolin.

Keywords

Coleus forskohlii · Forskolin · Endophytes *Fusarium redolens* · *Phialemoniopsis cornearis* · *Macrophomina pseudophaseolina* · *Piriformospora indica*

9.1 Introduction

Coleus forskohlii is a perennial medicinal shrub cultivated mainly for its forskolin content. The plant has been used since ancient times in ayurvedic traditional medicines for the treatment of hypertension, glaucoma, asthma, congestive heart failure, obesity, and cancer. The growth rhythm of the medicinal plant is slow and the alkaloid accumulation pattern is highly influenced by environmental and/or geographical conditions. Also, the quantity of the main secondary metabolite forskolin in natural conditions is found to be usually very low which restrains its commercial value. Plant–fungi interaction renders enhanced production of plant secondary metabolites as most of these compounds are produced due to activation of defense-related biosynthetic pathways. The use of endophytic microorganisms presents a special interest in the development of value-added bioactive compounds through agriculture. Limited investigations have been undertaken on in planta enhancement of forskolin content using endophytic fungus in sustainable agriculture.

T. Pullaiah, *Forskolin*, https://doi.org/10.1007/978-981-19-6521-0_9

Endophytes are an endosymbiotic group of microorganisms that reside in various tissues of plants without triggering any visible external sign of infection. Endophytes provide benefits to host plants and the environment. Interestingly, medicinal plants have been identified as a good host for a variety of endophytic microorganisms including fungi which synthesize secondary metabolites with biological activity (Pullaiah and Anuradha 2020). Plant–endophyte interactions can interfere with plant growth, development, and resistance against various stresses. Plant–fungi interaction renders enhanced production of plant secondary metabolites as most of these compounds are produced due to activation of defense-related biosynthetic pathways.

9.2 Endophytes and Forskolin Production

C. forskohlii being succulent in nature responds well to in vitro propagation and thus various explants viz., nodal segments, shoot tip, and leaf, are effectively used. Hence, symbiotic interaction and in vitro propagation in combination provide a promising alternative strategy to enhance accumulation of phytochemicals. *Piriformospora indica*, an endophytic root colonizing basidiomycete fungus, mimics Arbuscular Mycorrhizal Fungi (AMF) in many morphological, functional, and growth promotional aspects.

Piriformospora indica can colonize monocot as well as dicot plants and it acts as a bioregulator, biofertilizer, and bioprotector against root pathogens; it overcomes water stress (dehydration), acidity, desiccation, and heavy metal toxicity; it protects from pests, delays the wilting of the leaves, prolongs aging of callus tissues, it enhances secondary metabolites production, and also increases the nutritional value of the plant (Das et al. 2012). In contrast to AMF, *P. indica* can be easily grown on synthetic media and thus therefore can be very useful in sustainable agriculture for crop improvement. Das et al. (2012) investigated the influence of plant probiotic fungus *Piriformospora indica* on the medicinal plant *C. forskohlii*. Interaction of the *C. forskohlii* with the root endophyte *P. indica* under field conditions resulted in an overall increase in aerial biomass, chlorophyll contents, and phosphorus acquisition. Increases of 41% in length of the branches and of 44% in the number of leaves were observed in *P. indica* colonized plants as compared with the non-colonized plants. *P. indica* not only induced a faster development of the aerial part of the plant but also caused early maturation with respect to flowering. The faster development of *P. indica* colonized roots compared with the non-colonized plants during all stages of growth could be due to the earlier expression of developmentally regulated genes. The fungus also promoted inflorescence development, consequently, the amount of *p*-cymene in the inflorescence increased. Growth of the root thickness was reduced in *P. indica* treated plants as they became fibrous, but developed more lateral roots. Because of the smaller root biomass, the content of forskolin was decreased. The symbiotic interaction of *C. forskohlii* with *P. indica* under field conditions promoted biomass production of the aerial parts of the plant including flower development (Fig. 9.1). The plant's aerial parts are important source of metabolites for medicinal

Fig. 9.1 (**a**) Influence of *Piriformospora indica* on 6 months old *C. forskohlii* under field conditions. Top panel represents plant morphology as a result of interaction between *P. indica* on *C. forskohlii* under field conditions. Each polythene bag contained 2.5 kg of unsterile sand, field soil, and compost (1:1:0.25 w/w). The fungal inoculum was 2%(w/v). Each bag contained 30 days old rooted plant cuttings. Irrigation was done on alternate days using underground water. Lower panel represents root morphology of *C. forskohlii*. (**b**) Inflorescence in *C. forskohlii*. –Pi: non-colonized plants; +Pi: colonized plants. Photographs were taken after 6 months (Source Das et al. 2012)

application. Therefore, authors suggested that the use of the root endophyte fungus *P. indica* in sustainable agriculture will enhance medicinally important chemical production.

Das et al. (2014) conducted for optimization of in vitro substrates under aseptic conditions for interaction of *Piriformospora indica* with the medicinal plant *C. forskohlii*. They tested the effects of different substrates on *P. indica* colonization as well as growth parameters of the in vitro raised *C. forskohlii*. Interaction of in vitro *C. forskohlii* with root endophyte *P. indica* under aseptic conditions resulted in increase in growth parameters in fungus colonized plants. It was observed that *P. indica* promoted the plant's growth in all irrespective of substrates used for co-culture study. The growth was found inferior in liquid compared to semisolid medium as well as there was problem of hyperhydricity in liquid medium. *P. indica* treated in vitro plantlets were better adapted for establishment under greenhouse compared to the non-treated plants due to fungal intervention.

Deployment of plant endophytes at field level is reported to make an impact on agricultural crop productivity; development and deployment of suitable crop-specific plant probiotics in a suitable delivery matrix is a value-added task. Mastan et al. (2019b) attempted to develop bioformulations of native, fungal endophytes of *Coleus forskohlii* to improve plant yield using two different carrier-based materials (talc and wheat bran). Initially, fungal endophytes *Fusarium redolens* (RF1), *Phialemoniopsis cornearis* (SF1), and *Macrophomina pseudophaseolina* (SF2), were grown on sterilized wheat bran under solid-state condition and their growth kinetics and pattern were analyzed by ergosterol content and scanning electron microscope, respectively. Ten-day grown fungal endophytic cultures were used for the development of two types of formulations (wheat bran and talc-based formulations) and tested for their efficacy on host plant, *C. forskohlii* under field conditions. Interestingly, application of wheat bran-based endophytic formulations significantly enhanced plant height (12–29%), number of branches (51–63%), root biomass (26–33%), photosynthetic pigments (32–101%), and forskolin content (35–56%) compared to talc-based formulations under field conditions. Shelf life of endophytes (RF1, SF1, and SF2) in both formulations revealed spore viability in wheat bran-based formulations for 6 months storage period as compared to talc-based formulations. Overall, this investigation envisages developing plant probiotic bioformulations of functional endophytes of *C. forskohlii* to enhance root biomass and in planta forskolin content.

Mastan et al. (2019a) reported specific roles of three fungal endophytes, *Fusarium redolens* (RF1), *Phialemoniopsis cornearis* (SF1), and *Macrophomina pseudophaseolina* (SF2), functionally acting as plant probiotic fungus, regulating secondary metabolite (forskolin) biosynthesis in *C. forskohlii*. The root endophyte, RF1, and shoot endophytes, SF1 and SF2, were found to enhance forskolin content by 52–88% in pot and 60–84% in field experiments as compared to uninoculated control plants. The three endophytes also enhanced total biomass owing to plant growth-promoting properties. The expression of diterpene synthases (CfTPSs) like CfTPS1, CfTPS2, CfTPS3, and CfTPS4 were significantly upregulated in endophyte-treated *C. forskohlii* plants. Elevated expression of key diterpene synthases (CfTPS2) in the forskolin biosynthesis pathway, exclusively present in the root cork of *C. forskohlii*, was observed following SF2 endophyte treatment. Furthermore, endophyte treatments conferred a variety of antagonistic activity against nematode galls (80%) and plant pathogens like *Fusarium oxysporum, Colletotricum gloeosporioides*, and *Sclerotium rolfsii*. RF1 and SF1 fungal endophytes showed positive for IAA production; however, SF1 also indicated phosphate solubilization activity. Overall, the qualitative and quantitative improvement of in planta forskolin represents an area of high commercial interest, and hence, Mastan et al. (2019a) focused on novel insights for the application of three fungal endophytes for in planta enhancement of forskolin content for *C. forskohlii* cultivation by a sustainable approach.

Mastan et al. (2020) reported the role of plant–probiotic bacterial endophytes of *C. forskohlii*, CFLB1 and CFRB1, isolated from leaf and root, which regulate plant growth and in plant forskolin content. Native bacterial endophyte, CFRB1 (*Alcaligenes faecalis*), significantly modulated primary plant productivity and forskolin content under pot and field conditions. Under field conditions, CFRB1 endophyte application significantly enhanced photosynthetic pigments and reduced the severity of root-knot and root rot diseases. Expression analyses of functional genes involved in the forskolin biosynthesis in *C. forskohlii* plants treated with CFRB1 endophyte under field conditions revealed differential upregulation of four *C. forskohlii* diterpene synthases (CfTPSs), CfTPS1, CfTPS2, CfTPS3, and CfTPS4, along with cytochrome P450 (CfCYP76AH15) and acyltransferase (CfACT1-8) genes. CFRB1 treatment reduced the severity of nematode infection and root rot in *C. forskohlii* plants by 81 and 78%, respectively. Overall, cross-talk of plant–endophyte interaction in *C. forskohlii* is beneficial, leading to enhanced forskolin content through modulation of forskolin biosynthetic pathway genes along with increased plant yield and reduced disease incidence. Thus, endophytic isolate, *A. faecalis* (CFRB1) could be deployed as a novel bio-stimulant for enhancing in planta forskolin content during cultivation of *C. forskohlii.*

Field experiments were conducted by Mastan et al. (2021) to understand the compatibility of three native, endophytic fungi *Phialemoniopsis cornearis* (SF1), *Macrophomina pseudophaseolina* (SF2), and *Fusarium redolens* (RF1) with *Trichoderma viride* (TV1) on *Coleus forskohlii* in enhancing plant growth and forskolin content. Co-inoculation of RF1+TV1 showed significant improvement in plant growth (52%), root biomass (67%), and in planta forskolin content (94%), followed by treatment of SF2+TV1 and SF1+TV1. qRTPCR was carried out to quantify the expression of five key forskolin biosynthetic pathway genes (CfTPS2, CfTPS3, CfTPS4, CfCYP76AH15, and CfACT1-8) in RF1+TV1-treated *C. forskohlii* plants. Elevated expression of CfTPS2, CfTPS4, CfCYP76AH15, and CfACT1-8 genes was noticed with RF1+TV1 combination as compared to uninoculated *C. forskohlii* plants. Besides, RF1+TV1 treatment considerably reduced the severity of nematode infection of *C. forskohlii* plants under field conditions. Thus, congruent properties of *F. redolens* (RF1) were noticed with co-inoculation of *T. viride* (TV1) under field conditions resulted in enhanced forskolin content, root biomass, and reduced nematode infections in *C. forskohlii.*

Molecular identification of endophytic fungi associated with *C. forskohlii* was carried out by Crasta and Raveesha (2021). A total of 85 endophytic fungi were isolated from 280 leaf segments. Molecular identification revealed 34 fungal genera. Among these, species of *Cladosporium* sp., *Alternaria* sp., *Aspergillus niger, Aspergillus* sp., *Colletotrichum* sp., *Nigrospora oryzae, Penicillium* sp., and *Phyllosticta fallopiae* were found to be predominant genera. The percentage occurrence of members of Ascomycota was the highest, with 96.47% distribution and Basidiomycota members were distributed the least, with 3.53%.

References

Crasta GL, Raveesha KA (2021) Molecular identification of endophytic fungi associated with *Coleus forskohlii* (Willd.) Briq. J Appl Biol Biotechnol 9(6):162–172

Das A, Kamal S, Shakil NA, Sheramethi I, Oelmüller R, Dua M, Tuteja N, Johri AK, Varma A (2012) The root endophyte fungus *Piriformospora indica* leads to early flowering, higher biomass and altered secondary metabolites of the medicinal plant, *Coleus forskohlii*. Plant Signal Behav 7(1):103–112

Das A, Tripathi S, Varma A (2014) In vitro plant development and root colonization of *Coleus forskohlii* by *Piriformospora indica*. World J Microbil Biol Biotechnol 30(3):1075–1084

Mastan A, Bharadwaj R, Kushwaha RK, Babu CSV (2019a) Functional fungal endophytes in *Coleus forskohlii* regulate labdane diterpene biosynthesis for elevated forskolin accumulation in roots. Microb Ecol 78(4):914–926

Mastan A, Rane D, Dastager SG, Viviek Babu CS (2019b) Development of low-cost plant probiotic formulations of functional endophytes for sustainable cultivation of *Coleus forskohlii*. Microbiol Res 227:126310. https://doi.org/10.1016/j.micres.2019.126310

Mastan A, Rane D, Dastager SG, Vivek Babu CS (2020) Plant probiotic bacterial endophyte, *Alcaligenes faecalis*, modulates plant growth and forskolin biosynthesis in *Coleus forskohlii*. Probiotics Antimicrob Proteins 12(2):481–493. https://doi.org/10.1007/s12602-019-09582-1

Mastan A, Rane D, Dastager SG, Viviek Babu CS (2021) Molecular insights of fungal endophyte co-inoculation 1 with *Trichoderma viride* for the augmentation of forskolin biosynthesis in *Coleus forskohlii*. Phytochemistry 184:112654. https://doi.org/10.1016/j.phytochem.2021.112654

Pullaiah T, Anuradha M (2020) Endophytes and Camptothecin production. In: Pullaiah T, Raghavendra PB, Karuppusamy S, Raveendran V, Anuradha M (eds) Camptothecin and Camptothecin producing plants. Elsevier, pp 317–338

Propagation Methods of *Coleus forskohlii* 10

Abstract

In the present chapter, propagation methods of *Coleus forskohlii* are given. It is propagated by seeds as well as vegetatively by terminal stem cuttings. The crop is cultivated through 55-day-old rooted cuttings which are generally planted on ridges at a spacing of 60 × 45 cm. Soil conditions, climate requirements, sowing time, crop duration, planting material, spacing, and economics of cultivation are given. Organic and inorganic fertilizer application dosages are also provided. Diseases and pests of *Coleus* crop and their management are also given.

Keywords

Coleus forskohlii · Propagation · Forskolin

10.1 Introduction

Indiscriminate collection of *Coleus forskohlii,* the only source of the forskolin, from the wild has made the species vulnerable (Vishwakarma et al. 1988). In the year 2015–2016, India exported a total of 29.69 tons of *C. forskohlii* products to different countries worth 2.62 million USD (https://www.eximpulse.com/export-product-Coleus-Root.htm?tpages=3). As collection from wild sources is detrimental to the existence of this species there is a necessity for the development of propagation techniques.

10.2 Propagation

Coleus crops need irrigation and can be sown at any time. Sandy loamy soil is good for this crop. *C. forskohlii*, in its native environment, grows on loamy or sandy loam. *Coleus* performs best on well-drained soils. Balasubramanian et al. (2020) evaluated

T. Pullaiah, *Forskolin*, https://doi.org/10.1007/978-981-19-6521-0_10

soil characteristics and yield variation of *C. forskohlii* in different agroclimatic zones of Tamil Nadu.

Relative humidity between 83% and 95% and a temperature of 10–25 °C is ideal for the crop. Annual rainfall of 100–160 cm is required (Shah and Kalakoti 1996). Coleus is propagated by seeds as well as vegetatively by terminal stem cuttings. Seed propagation is difficult and slow whereas propagation by terminal stem cutting is easy and economical. 10–12 cm long terminal cuttings with 3–4 pairs of leaves are planted in nursery beds to induce rooting (Nisar et al. 2020). One-month-old stem cuttings which have sufficient roots are transplanted to the main field. The best period for planting is during June/July and September/October (Rajamani and Vadivel 2009). Organic manure is required to the level of 140 kg on the 30th day and 45th day of planting. A combination of 40 kg N, 60 kg P_2O_5, and 50 kg K_2O per ha is optimum for obtaining the maximum fresh (120 t/ha) and dry (3.982 t/ha) tuber yield (Veeraraghavathatham et al. 1985).

Patel (2016) studied vegetative propagation of *C. forskohlii* using their mature stem cuttings in poly bags including sand, manure, and soil mixture equally. Selected stem cuttings for the above purpose were allowed to grow in each one of the poly bags with a medium water supply at the starting of the experimentation. After development of the new plants, these were carefully removed from poly bags and were further grown in the prepared beds in herbal garden to multiply (Figs. 10.1, 10.2, 10.3, 10.4, 10.5, and 10.6).

Coleus plants raised in presence of the arbuscular mycorrhizal fungi *Glomus bagyarajii* showed an increase in plant growth and forskolin content over those grown in the absence of AM fungi (Sailo and Bagyaraj 2005). Synergy between *Glomus fasciculatum* and a beneficial *Pseudomonas* in reducing root diseases and improving yield and forskolin content in *C. forskohlii* under organic field conditions has been studied by Singh et al. (2013).

Damam et al. (2014) investigated the effect of three plant growth-promoting rhizobacterial strains *Pantoea* sp. and *Pseudomonas* sp. in individual or in combination treatments on *C. forskohlii*.

In cultivation, the size of *C. forskohlii* plants varies, but these plants can grow several feet tall and wide. Malek et al. (2019) investigated the effect of different cutting types and IBA concentration on survival and rooting of the cuttings. Growth attributes show variation with different cutting types and IBA. Rooting percentage and survival was 100 percent in all treatments at 45 DAP. Stem cuttings treated with brassinosteroids gave more root formation and root growth over control (Swamy and Rao 2010). 28-homobrassinolide at 3 μM concentration was highly effective in enhancing the growth. The maximum sprouting, node sprouting, and survival for cuttings treated with 100 ppm IBA (Tiwari and Das 2010). Sundharaiya et al. (2000) also confirmed the terminal cuttings treated with 500 ppm IBA recorded the highest rooting percentage. The temperature clearly affected the production of forskolin, demonstrating that the contents of forskolin were higher at 20 °C than at 15 and 30 °C (Yanagihara et al. 1995).

Fifty-five-day-old, rooted cuttings are generally planted on ridges at a spacing of 60 cm (row to row) and 45 cm (plant to plant) (Fig. 10.1). Mastiholi et al. (2010)

Fig. 10.1 *C. forskohlii* cuttings planting in poly bags (Source: Patel 2016)

reported that closer spacing of 60 cm × 20 cm recorded significantly higher gross and net returns and B: C ratio (3.25) than wider spacing levels (60 cm × 30 cm, 75 cm × 20 cm and 45 cm × 30 cm). The percent increase in net returns at 60 cm × 20 cm spacing was 98% over 60 cm × 30 cm. Higher tuber yield recorded at closer spacing gave higher returns. As the spacing was reduced from 60 cm × 30 cm to 60 cm × 20 cm there was a corresponding increase in tuber yield per unit area which was mainly attributed to increase in plant population (55,555 to 83,333 plants/

Fig. 10.2 *C. forskohlii* growth in polybags after 10 days (Source: Patel 2016)

ha, respectively). The increased tuber yield due to a decrease in spacing was manifested in increase in net returns (Rs. 28,697–56,822/ha, respectively). Thus, the higher returns at closer spacing were due to higher plant population (51%). Under irrigated conditions, production per unit area rather than individual plant performance decides the profitability of the crop (Mastiholi et al. 2010). Higher net returns and B: C ratio at closer spacing were also reported by Jayalakshmi (2003).

Fig. 10.3 *C. forskohlii* growth in polybags after 20 days (Source: Patel 2016)

Veeraraghavathatham et al. (1985) stated that when the plant population was increased there was a gradual increase in the tuberous root yield per unit area and it was highest in the closest planting pattern of 60X20 cm in *C. forskohlii.*

Studies by Mastiholi et al. (2013a, b) showed higher total biomass (TBP) when planted at a spacing of 60 cm × 20 cm followed by75 cm × 20. There was an increase in tuber biomass production with a decrease in plant-to-plant spacing (1.04 t/ha at 60 cm × 30 cm and 1.57 t/ha at 60 cm × 20 cm).

Transplanting is done into planting holes having a depth of 10–12 cm and dia. of 8–10 cm. On average, an yield of 800–1000 kg/ha of dry tubers may be obtained. However, if proper cultivation practices are applied, an yield of up to 2000–2200 kg/ha of dry tubers can easily be obtained (Rajamani and Vadivel 2009).

Prabhaharan (2005) studied the effect of organics and inorganics on the yield and quality of *C. forskohlii* in the alfisols of southern Tamil Nadu. Mastiholi (2008) reported that there was a gradual increase in plant length with increase in NPK doses. Jayalakshmi (2003) found a significant influence in growth of coleus with the application of increased phosphorus levels. Nadukeri et al. (2014) reported that application of 75 percent RDF (Recommended dose of fertilizer) + 10 t FYM + vermicompost 5 t per hectare produced increased plant height (66.49 cm), number of branches per plant (85.95), leaf area index (7.49) at harvest, absolute growth rate (3.394 g/plant/day), crop growth rate (0.943 g/m^2/day), and relative growth rate (0.0460 g/g/week) were recorded at 120–160 days after planting.

Saraswati et al. (2016, 2019) investigated the effect of organic manures on growth, yield, quality, and nutrient uptake of *C. forskohlii*. Application of pongamia cake equivalent to 100 percent RDN (Recommended dose of nitrogen) recorded the highest plant height and number of branches. Plants provided with FYM equivalent to 100 percent RDN gave 1.28 t per hectare of dry tuberous root yield. However, poultry manure, pongamia cake, pressmud, and RDF application also gave on par dry tuberous root yield.

Nadukeri and Kattimani (2012) investigated the integrated nutrient management in *C. forskohlii* and concluded that growing coleus with application of 75%

Fig. 10.4 *C. forskohlii* growth in polybags after 30 days (Source: Patel 2016)

RDF + 10 t FYM + vermicompost 2.5 t/ha or 75% RDF + 10 t FYM + vermicompost 5.0 t /a had beneficial effects on obtaining maximum net returns and cost-benefit ratio. Influence of FYM, inorganic fertilizer (NPK), and sources of potassium on the yield of *C. forskohlii* was investigated by Somanath (2002) and Somanath et al. (2005).

The effect of organic manures (different doses of castor cake, FYM), biofertilizer (*Azospirillum*, phosphorous solubilizing bacteria), and inorganic nitrogenous fertilizer (different levels) on tuber yield at harvest and soil nutrient status at different growth stages (90, 120 DAP and at harvest) were investigated by Sandya Rani et al. (2009).

Priya et al. (2013) assessed the role of organic manures, biofertilizers, and their interaction effect on the growth and yield of *Coleus*. FYM 5 t + vermicompost 0.5 t/ha among organic manures and Azospirillum + Phosphobacter each @ 2 kg/ha

Fig. 10.5 *C. forskohlii* planting in the field (Source: Patel 2016)

among biofertilizers significantly increased the growth and yield and was on par with inorganic fertilization. The effects of farmyard manure (FYM) and NPK fertilizers on the productivity of *C. forskohlii* were studied in Bangalore, Karnataka, India, during the Kharif of 2002 by Somanath et al. 2005

Fig. 10.6 Planting of *Coleus forskohlii* rooted cuttings (Source: https://agritech.tnau.ac.in/horticulture/horti_medicinal%20crops_medicinal%20coleus.html)

Under irrigated conditions, harvesting the crop at the physiological maturity stage (Fig. 10.2) helps to obtain higher yields and returns. Better utilization of the growing season by planting succeeding crop also warrants harvesting the preceding crop at the right time. Jayalakshmi (2003) recorded higher net returns at 180 DAP. Mastiholi et al. (2010) reported that a higher B: C ratio and returns were obtained by harvesting at 180 and 160 days after planting (DAP). The higher returns at the above harvesting stage were due to higher tuber yield. The percent increase in net returns at 180 DAP over 160 DAP was only 5.4%. But, there was a 77.2 percent increase in net returns at 160 DAP over 140 DAP. The increase in B: C ratio at 180 DAP over 160 DAP (3.2%) was also negligible, compared to 41.5% increase at 160 DAP over 140 DAP. Mastiholi et al. (2010) concluded that under irrigated condition, cultivation of *C. forskohlii* becomes profitable when it is planted at the spacing of 60 cm × 20 cm and harvested at 160 DAP. By advancing the harvesting time by 20 days (harvesting at 160 DAP instead of 180 DAP) farmers can suitably accommodate succeeding crops under irrigated conditions in addition to getting optimum yield and returns. Mastiholi et al. (2010) investigated the effect of NPK levels on forskolin content and yield in *C. forskohlii* under irrigated conditions.

Nageswara Rao et al. (2011) reported that 50% of NPK and 10 tonnes of farm yard manure per hectare at the harvesting time of 150 days are ideal for the maximum yield of forskolin. Field experiments were carried out by Vennila and

Jayanthi (2014b) on sandy loam soil to assess the effect of planting systems (normal planting and paired row planting) and sources of nutrients on *C. forskohlii.*

A 2 years field experiment carried out by Vennila and Jayanthi (2014c) indicated that application of 40:60:50 kg NPK/ha + 10 t FYM/ha or application of poultry manures at 3 t/ha is liable for producing higher yield, nutrients uptake, and nitrogen use efficiency. Nageswara Rao (2014) also reported that combination of organic and inorganic fertilizers is very productive than the application of manures or fertilizers alone for achieving higher growth and yield in *C. forskohlii.* An experiment was carried out by Sathiyaraj (2017) on integrated nutrient management combination with biofertilizers and plant growth substances on yield and quality of *C. forskohlii.*

An experiment was conducted by Muruganandam et al. (2021) on nutrient management of *C. forskohlii* with different combinations of nutrients. Among the various treatments tried, plants supplied with T5 (75% RDF + FYM @ 15 t/ha + Castor cake at 2 t/ha + Azotobacter at 10 kg/ha) was recorded maximum growth parameters.

Shukla et al. (2022) investigated chemotypic variability of seven germplasm samples of *C. forskohlii* collected from different phytogeographical locations in India. Data on soil analysis correlated with the bioactive compounds. Quantification of forskolin and iso-forskolin revealed a wide range of variations, varying from 1.15 to 0.004% and 0.0091 to 0.1077% per dry weight basis, respectively.

After 4.5–5 months of planting the crop of *C. forskohlii* gets ready for harvesting. The plants are uprooted, tubers separated, cleaned, and sun dried (Figs. 10.7 and 10.8). The fresh root tubers just after harvesting contain 75–85% moisture level, which goes down to 12% due to drying. Root tubers can be dried by two methods: Sun drying and mechanical drying. Mechanical drying requires a temperature of 40 ° C to dry the root tuber. After drying, root tubers having a slice thickness of 0.5 cm

Fig. 10.7 *Coleus forskohlii* in cultivation (Source: https://www.nutraceuticalsworld.com/contents/view_breaking-news/2019-04-04/sabinsa-obtains-patents-for-forslean-coleus-forskohlii-extract-related-to-energy-balance/)

Fig. 10.8 Cleaning of harvested tubers of *Coleus forskohlii* (Source: https://agritech.tnau.ac.in/horticulture/horti_medicinal%20crops_medicinal%20coleus.html)

are packed in a bag that is lined with polyethylene. Mechanical drying retains a high yield of forskolin as compared to Sun drying. Variety CO.1 Medicinal coleus (*C. forskohlii*), selected from Theni local, recorded 1.98 tonnes dry tuber/ha with forskolin content of 0.54%. Moderately resistant to root rot and wilt diseases under field conditions. Gupta et al. (2021) determined the effect of Plant Growth Regulators (PGRs) viz. IAA, GA_3, and Kinetin in their physiological ranges from 10^{-7} to 10^{-5} M on tuberous roots of *C. barbatus* with their respective control.

C. forskohlii plant was subjected by Khan et al. 2020 to different water stress conditions under controlled temperatures in climate control Greenhouses. The forskolin and proline contents were found to be maximum in the roots of plants with 80% moisture stress. Similarly, the potassium content was also influenced by the water stress treatment. However, the chlorophyll content was found to be maximum in plants with 20% moisture stress. The soil moisture stress has a significant influence on forskolin and proline content. Hence, the crop may be cultivated in a stress condition for maximum production of forskolin.

There are several other publications on the propagation of *Coleus forskohlii* and these include Farooqi and Sreeramu (2004), Sudhakar (2005), Srinivas et al. (2008), Somanath et al. (2004), Mastiholi (2010) and Priya et al. (2013). Several PhD and MSc theses were submitted on the propagation of *C. forskohlii*, these include

Veeraraghavathatham et al. (1988), Patil (2000), Sailaja (2004), Rajangam (2005), Sudhakar (2005), and Vennila (2006). From the above review, it is clear that almost all the propagation studies were carried out in India.

10.3 Intercropping

Field experiments were carried out by Vennila (2006), Vennila et al. 2008, Vennila and Jayanthi 2014a) on sandy loam soil to assess the effect of planting systems and sources of nutrients on *C. forskohlii* + *Phyllanthus amarus* intercropping system. Planting of *C. forskohlii* and *P. amarus* (1:1 ratio) at a spacing of 90–30 × 30 cm registered higher tuber yield. Effect of spacing, manure, and fertilizers on the growth and yield of *C. forskohlii* under teak plantation was investigated by Suresh et al. (2010).

Cultivation and intercropping trial of *C. forskohlii* for economic yield from roots and whole plant were calculated under sapota-jatropha plantations by Malek et al. (2020) in south Gujarat, India. The economic feasibility of intercropping of coleus crop with sapota-jatropha was calculated higher under sapota + jatropha based horti-medicinal agroforestry system. The primary results for intercropping of *C. forskohlii* can also be replicated with other horti-silvi systems viz., mango–bamboo, mango–teak, and sapota–gmelina for better economic returns to the farmers.

10.4 Diseases

Diseases and their management in *C. forskohlii* have been reviewed by Singh et al. (2011) and these are briefly mentioned below.

10.4.1 Leaf Spot Disease

Leaf spot disease in *C. forskohlii* is caused by fungus *Corynespora cassiicola* (Fernandes and Barreto 2003) and *Botryodiplodia theobromae* (Ramprasad 2005).

10.4.2 Blight Disease

Shukla et al. (1993) reported that *Rhizoctonia solani* causes leaf blight in *C. forskohlii*.

10.4.3 Root Rot/Wilt Disease

Fusarium chlamydosporum causes root rot disease in *C. forskohlii* (Shyla 1998; Singh et al. 2009). *F. solani* causing root rot of *C. forskohlii* has also been reported by Bhattacharya and Bhattacharya (2008). *Ralstonia solanacearum* was reported to be causing the vascular wilt of *C. forskohlii* (Coelho and Assis 2002; Chandrashekara and Prasannakumar 2010).

Trichoderma harzianum and zinc sulfate exerted maximum reduction in root rot incidence caused by *Macrophomina phaseolina* (Kamalakannan et al. 2006). Muthulakshmi et al. (2021) investigated the effect of bioagents on the management of root rot disease in *Coleus*. They reported that basal application of *Bacillus subtilis* (Bbv 57) at 2.5 kg/ha + dipping root cutting in 2% (Bbv 57) + soil application of Bbv 57 at 2.5 kg/ha for 90 days after planting followed by treatment with T1 with basal soil application of *B. subtilis* (Bs1) at 2.5 kg/ha + dipping cuttings in 0.2% *B. subtilis* (Bs1) + soil application of *B. subtilis* (Bs1) 22.5 kg/ha was significantly superior in reducing the root rot incidence and increasing the tuber root yield.

10.4.4 Root Knot Disease of *C. forskohlii*

Root knot disease in *C. forskohlii* has been reported to be caused by *Meloidogyne incognita* and *M. arenaria. M. incognita* has been reported to cause a yield reduction of up to 86% (Senthamarai et al. 2006), while severe losses also occur in *C. forskohlii* because of *M. arenaria* infestations (Bhandari et al. 2007). A pot culture experiment was carried out by Senthamarai et al. (2008) under glasshouse conditions for the evaluation of biocontrol agents against *M. incognita* in *C. forskohlii.*

10.4.5 Complex Disease of *C. forskohlii*

Collar rot complex of *C. forskohlii* involving *Fusarium chlaydosporum* and *Rhizoctonia bataticola* (*Macrphomonia phaseolina*) was reported by Kulkarni et al. (2007). Complex disease of *C. forskohlii* has also been reported involving both fungal and nematode pathogens (Senthamarai et al. 2008).

10.4.6 Stem Blight

Ramprasad (2005) reported stem blight caused *Phytophthora nicotianae* var. *nicotianae.*

10.5 Disease Management

Water stagnation in *C. forskohlii* fields may lead to severe infections of *Fusarium* and *Ralstonia,* therefore, water stagnation in the planted fields should be avoided.

10.5.1 Chemical Control

10.5.1.1 Fusarial/Bacterial Wilt Control

Dipping the terminal cuttings in carbendazim solution (1 g/l) before planting protects *C. forskohlii* from fungal pathogens. The chemical Emisan (0.2%) has been found to protect the plants against *Fusarium* wilt to some extent but the protection provided to plants inoculated with biocontrol agents *Trichoderma viride* and *Glomus mosseae* was found to be higher (Boby and Bagyaraj 2003). *Glomus fasciculatum* and *Pseudomonas fluorescens* are most effective treatments that reduced 56–65% and 61–66%, under lower and higher levels of pathogen *Fusarium chlamydosporum* (Singh et al. 2009). Chemical fungicides (benomyl) reduced the disease incidence (54.54%) caused by *F. chlamydosporum* (Singh et al. 2009) during a field study of *C. forskohlii*. Paramasivan et al. (2007) reported that the use of chemical fungicide (Carbendazim) reduced the disease incidence by 18%. Kulkarni et al. (2007) reported that the lowest population (cfu/g soil) of *F. chlamydosporum* and *R. bataticola* was observed with the use of carbendazim (Singh et al. 2011).

10.5.2 Biological Control

10.5.2.1 Fusarial/Bacterial Control

Arbuscular mycorrhizal (AM) fungi suppressing the activity of root pathogens are well documented (Mohan and Verma 1996). *P. fluorescens*, mainly considered as a PGPR, can suppress a wide range of plant pathogens including *Fusarium* (Defago and Hass 1990). Neem and neem products are effective against the root/soil-borne pathogens (Singh et al. 2011).

Paramasivan et al. (2007) reported that the use of bioinoculants like *T. viride* and *P. fluorescens* reduced the disease incidence by 20–21%. A combination of *T. viride* + Neemato (neem-based product applied at 500 g/5 m^2) resulted in the lowest wilt incidence by 12.76% (Kulkarni et al. 2007).

Two-year field experiments were conducted by Singh et al. (2012) with five bioinoculants and neem cake under organic field conditions (with vermicompost as a nutritional supplement) to evaluate their potential to control root rot and wilt (a complex problem involving *Fusarium chlamydosporum* and *Ralstonia solanacearum*) of *C. forskohlii*.

10.6 Mulching and Weed Control

The effect of mulching with organic materials and black polythene sheet on weed population and tuber yield of *C. forskohlli* was evaluated by Gunasekaran and Shakila (2014).

10.7 Crop Improvement

Crop improvement studies were reviewed in forskolin-yielding *C. forskohlii* by Hegde and Kumar (2002). Wide variability was reported in morpho-economical characters in available germplasm. Among tuberous and non-tuberous accessions, a cultivated tuberous type "K" recorded higher tuber yield as well as higher forskolin content in dry tubers. Velmurugan et al. (2009) studied correlation and path analysis in mutants of *C. forskohlii* for yield and forskolin content in V2M1 generation.

A field evaluation trial was conducted by Hegde et al. (2005) involving 13 accessions of *C. forskohlii*. The accession IIHR-80 was with medium tuber yield and higher forskolin content (0.715%). Sharma and Vasundhara (2015) evaluated the *C. forskohlii* varieties and mutants against the local check (K-8) under sterilized media (cocopeat) and nutrient solution (864 mg plant^{-1} nitrogen, 768 mg plant^{-1} phosphorous and 960 mg plant^{-1} potassium) for growth, yield, and forskolin content. Mutant MV2 was found to be better than other varieties.

References

Balasubramanian V, Suresh J, Srinivasan R, Prabhu M, Manikandan E, Gokul Nandha VB (2020) Evaluation of soil characteristics and yield variation of coleus (*Coleus forskohlii*) in different agro climatic zones of Tamil Nadu. Int J Chem Stud 8(4):2841–2845

Bhandari S, Harsh NSK, Singh P (2007) First report on *Meloidogyne arenaria* on *Coleus forskohlii* in India. Indian Forester 133:1709–1710

Bhattacharya A, Bhattacharya S (2008) A study on root rot disease of *Coleus forskohlii* Briq. Occurring in Gangetic West Bengal. J Bot Soc Bengal 62:43–47

Boby VU, Bagyaraj DJ (2003) Biological control of root-rot of *Coleus forskohlii* Briq. using microbial inoculants. World J Microbil Biotechno 19:175–180

Chandrashekara KN, Prasannakumar MK (2010) New host plants for *Ralstonia solanacearum* from India. New Dis Rep 22:6

Coelho N, Assis LAG (2002) *Coleus barbatus*: Um Novo Hospedeiro De *Ralstonia solanacearum*. Fitopatolo bras 27(2):1066

Damam M, Gaddam B, Kausar R (2014) Effect of plant growth promoting Rhizobacteria (PGPR) on *Coleus forskohlii*. Int J Curr Microbiol App Sci 3(9): 266–274

Defago G, Hass D (1990) *Pseudomonas* as antagonists against soil borne plant pathogen: mode of action and genetic analysis. Soil Biochem 6:249–291

Farooqi AA, Sreeramu BS (2004) Coleus. In: Cultivation of medicinal and aromatic crops. Universities Press, Hyderabad, pp 85–89

Fernandes RC, Barreto RW (2003) *Corynespora cassiicola* causing leaf spots on *Coleus barbatus*. Plant Patho 52:786

Gunasekaran P, Shakila A (2014) Effect of mulching on weed control and tuber yield of medicinal coleus (*Coleus forskholli* Briq.). Asian J Hort 9(1):124–127

Gupta A, Gupta RK, Dhingra GK, Kuriyal S, Lal S, Arya R (2021) Effect of plant growth regulators (PGRs) in their physiological ranges on tuberous roots of *Coleus barbatus*. Intern J Agric Sci 11(1):1–7

Hegde L, Kumar TV (2002) Genetic improvement in *Coleus forskohlii* Briq. Crop improvement, production technology, trade and Commerce, pp 97–105

Hegde L, Kumar TV, Himabindu K (2005) Evaluation of *Colus forskohlii* accessions for tuber and forskolin yield. Acta Hortic 675:217–219

https://www.eximpulse.com/export-product-Coleus-Root.htm?tpages=3
Jayalakshmi S (2003) Effect of spacing and nitrogen levels on growth, tuberous roots and yield and alkaloid content of medicinal coleus (*Coleus forskohlii* Briq). Thesis, Tamil Nadu Agricultural University, Coimbatore, India, M.Sc

Kamalakannan K, Mohan L, Samiyappan R, Chandrasekaran A, Valluvaparidasan V (2006) Development of inoculums dependent management strategies for the control of coleus (*Coleus forskohlii*) root rot caused by *Macrophomina phaseolina.* In: proceedings of international conference on Globalization of traditional, complementary and alternative system of medicine, held at TNAU, Coimbatore, p 45

Khan NA, Shaktawat GS, Upadhyaya SD, Upadhyay A (2020) Biochemical constituents of medicinal important *Coleus forskohlii* Briq. (Patharchur), under soil-moisture stress. Intern J Environ Health Sci 2(3):32–36

Kulkarni MS, Ramprasad S, Hedge Y, Laxminarayan H, Hedge NK (2007) Management of collar rot complex disease of *Coleus forskohlii* (Willd.) Briq. using bioagents, organic amendments and chemicals. Biomed 2:37–40

Malek SS, Dobriyal MJ, Desai BS (2019) Propagation of medicinal plant *Coleus forskohlii* (Poir.) Briq. (Pathachur or Galbel). Acta Scientifica. Agriculture 3(3):142–146

Malek SS, Dobriyal MJ, Thakur NS, Desai BS, Gunaga RP (2020) *Coleus forskohlii* (Poir.) Briq.: a suitable intercrop of horti-medicinal agroforestry system for South Gujarat farmers. Trends in Biosciences 13(3):162–167

Mastiholi AB (2008) Effect of spacing, time of harvest and NPK levels on growth, yield and quality of medicinal coleus (*Coleus forskohlii* Briq.) under irrigated condition. Ph. D. Thesis, University of Agricultural Sciences, Dharwad, India

Mastiholi AB (2010) Effect of NPK levels on forskolin content and yield in medicinal coleus (*Coleus forskohlii* Briq.) under irrigated condition. Green Farming 1(2):135–137

Mastiholi AB, Hiremath SM, Patil PL, Chittapur BM, Hegde L, Mathad JC (2010) Economics of pashanbhedi (*Coleus forskohlii* Briq.) as influenced by spacing, time of harvest and their interactions under irrigated condition. Indian. J Agric Res 44(3):221–224

Mastiholi AB, Prabhuling G, Bhairappanavar ST (2013a) Effect of spacing and time of harvest on nutrient uptake in medicinal coleus under irrigated condition. Asian J Hortic 8(2):596–600

Mastiholi AB, Chavan ML, Prabhuling G (2013b) Biomass production and its accumulation in *Coleus forskohlii* Briq.: impact of spacing and time of harvest. J Root Crops 39(2):131–135

Mohan V, Verma N (1996) Interaction of VA mycorrhizal fungi and rhizosphere microflora of forest tree species in arid and semi arid regions. Proceedings of IUFRO, In: An Impact of diseases and insect pests in tropical forests. pp. 222–223

Muruganandam C, Ezhilnilavu R, Sivasankar S (2021) Effect of integrated nutrient management on growth parameters of medicinal coleus (*Coleus forskohlii* Briq.). Plant Archives 21:2525–2528

Muthulakshmi P, Prabhakar K, Rajamani K (2021) Biomanagement of root rot in *Coleus forskohlii* caused by *Macrophomina phaseolina*. Medicinal Plants 13(2):318–321

Nadukeri S, Kattimani KN (2012) Integrated nutrient management in coleus (*Coleus forskohlii* Briq.). Karnataka J Agric Sci 25(3):407–408

Nadukeri S, Kattimani KN, Kolakar SS (2014) Influence of organic and inorganic fertilizers on growth and tuber yield of coleus (*Coleus forskohlii* Briq.) under northern dry zone of Karnataka. Intern J Agric Sci 10(1):119–123

Nageswara Rao S (2014) Integrated nutrient management for growth and high yield in *Coleus forskohlii*. Intl J Appl Biol Pharamaceut Technol 5(4):291–295

Nageswara Rao S, Gangadhar Rao S, Bhargavi B, Janardhan Reddy K (2011) Effect of NPK and farm yard manure on accumulation of forskolin, an antiobesity diterpenoid molecule in *Coleus forskohlii*. Curr Trends Biotechnol Pharm 5:1346–1352

Nisar S, Hanif MA, Soomro K, Jilani MI, Kala CP (2020) Chapter 11. Coleus. In: Hanif MA, Nawaz H, Khan MM, Byrne HJ (eds) Medicinal plants of South Asia: novel sources for drug discovery. Elsevier, pp 136–147

Paramasivan M, Mohan S, Muthukrishnan N (2007) Management of *Coleus* dry root rot pathogen *Macrphhomina phaseolina* by fungal and bacterial antagonist. Indian J Plant Prot 35:133–135

Patel DK (2016) Vegetative propagation of *Coleus forskohlii* (Willd.) Briq. using their stem cutting for ex-situ conservation in herbal garden. Med Aromatic Plants 5:4. https://doi.org/10.4172/2167-0412.1000261

Patil S (2000) Evaluation of genotypes and standardization of production technology of *Coleus forskohlii* Briq. PhD Thesis. UAS, Bangalore

Prabhaharan J (2005) Effect of organics and inorganics on the yield and quality of coleus (*Coleus forskohlii* Briq.) in the alfisols of southern Tamil Nadu. M.Sc. (Ag.) Thesis. Agricultural College and Research Institute, Killikulam. TNAU, Coimbatore

Priya BT, Rama Devi P, Sireesh K, Sunitha P (2013) Growth and yield of *Coleus forskohlii* as influenced by organic manures and bio-fertilizers. Plant Archives 13:1109–1112

Rajamani K, Vadivel E (2009) Marunthu Kurkan –Medicinal *Coleus*. In: Naveena Mulikai Sagupaddi Thozhil Nuttpangal. Tamil Nadu Agricultural University, Coimbatore, pp 17–22

Rajangam J (2005) Studies on the effect of planting methods and growth regulators on growth, tuber development, yield and quality, and standardization of postharvest technology of coleus (*Coleus forskohlii* Briq.). PhD Thesis. Tamil Nadu Agricultural University, Coimbatore

Ramprasad S (2005) Studies on collar rot complex of *Coleus forskohlii* (Wild.) Briq, M.Sc. Thesis, UAS, Dharwad

Sailaja J (2004) Effect of organic manures and inorganic nitrogenous fertilizer on growth, development and yield of medicinal coleus (*Coleus forskohlii* Briq.). M.Sc., (Hort.) Thesis, Acharya N.G. Ranga Agricultural University, Hyderabad

Sailo GL, Bagyaraj DJ (2005) Influence of different AM-fungi on the growth, nutrition and forskolin content of *Coleus forskohlii*. Mycol Res 109:795–798

Sandya Rani K, Devi MU, Patnaik MC, Kumar MR (2009) Integrated nutrient management on medicinal coleus and soil nutrient status. Crop Res 36:341–348

Saraswati SS, Mastiholi AB, Ajithkumar S, Chavan ML, Sakhubai HT (2016) Effect of organic manures on growth, yield, quality & nutrient uptake of medicinal coleus (*Plectranthus forskohlii* Willd.). Green Farming 6(6):1276–1280

Saraswati SS, Mastiholi AB, Srikantaprasad D, Ajithkumar S, Raghavendra S (2019) Yield and quality of medicinal coleus (*Plectranthus forskohlii* Willd.) cultivar 'K-8′ influenced by organic manures. Acta Hortic 1241:357–362

Sathiyaraj (2017) Integrated nutrient management combination with bio-fertilizers and plant growth substances on yield and quality of medicinal coleus (*Coleus forskohlii* Briq.). Int J Curr Microbiol App Sci 6(11):3329–3339

Senthamarai K, Poornima K, Subramanian S (2006) Pathogenicity of *Meloidogyne incognita* on *Coleus forskohlii* Briq. Indian J Nematol 36(1):123–125

Senthamarai K, Poornima K, Subramanian S, Sudheer J (2008) Nematode-fungal disease complex involving *Meloidogyne incognita* and *Macrophomina phaseolina* on medicinal Coleus, *Coleus forskohlii* Briq. Indian J Nematol 38(1):30–33

Shah V, Kalakoti BS (1996) Development of *Coleus forskohlii* as a medicinal crop. In: International Conference on Domestication and commercialization of non-timber forest products in agroforestry systems. FAO Corporate Document Repository

Sharma Y, Vasundhara M (2015) Influence of substrate culture on growth, yield and forskolin content of coleus (*Plectranthus barbatus* Andr.) varieties/mutants. Med Plants 7(2):131. https://doi.org/10.5958/0975-6892.2015.00020.9

Shukla PK, Misra A, Srivastava A, Kumar M, Srivastava S (2022) Study on chemotypic variability of *Coleus forskohlii* Briq., samples collected from different phytogeographical locations of India and evaluation of its inhibitory potential. J Chromatographic Sci:bmac033. https://doi.org/10.1093/chromsci/bmac033

Shukla RS, Kumar S, Singh HN, Singh KP (1993) First report of aerial blight of *Coleus forskohlii* caused by *Rhizoctonia solani* in India. Plant Dis 77:429

Shyla M (1998) Etiology and management of root-rot of *Coleus forskohlii*. M.Sc. Thesis, University of Agricultural Sciences, Bangalore

Singh R, Gangwar SP, Singh D, Singh R, Pandey R, Kalra A (2011) Medicinal plant *Coleus forskohlii* Briq.: disease and management. Med Plants 3(1):1–7

Singh R, Kalra A, Ravish BS, Divya S, Parameswaran TN, Srinivas KVNS, Bagyaraj DJ (2012) Effect of potential bio-inoculants and organic manures on root- rot and wilt, growth, yield and quality of organically grown *Coleus forskohlii* in a semi-arid tropical region of Bangalore (India). Plant Pathol 61:700–708

Singh R, Parameswaran TN, Divya S, Puttanna K, Satyasrinivas KVN, Bagyaraj DJ, Kalra A (2009) Management of root-rot/wilt of *Coleus forskohlii* Briq. In: CIMAP Golden Jubilee National Symposium on Medicinal & Aromatic Plants "Fifty Years of Research on Medicinal & Aromatic Plants". CIMAP, RC, Bangalore, p 18

Singh R, Soni SK, Kalra A (2013) Synergy between *Glomus fasciculatum* and a beneficial *Pseudomonas* in reducing root diseases and improving yield and forskolin content in *Coleus forskohlii* Briq. under organic field conditions. Mycorrhiza 23(1):35–44

Somanath S (2002) Effect of FYM, fertilizers (NPK) and sources of potassium on growth, yield and quality in *Coleus forskohlii* Briq. M.Sc. (Hort.) Thesis, University of Agricultural Sciences, Bangalore

Somanath S, Bhaskar S, Sreenivasmurhy CA (2005) Influence of FYM, inorganic fertilizer (NPK) and sources of potassium on the yield of *Coleus forskohlii*. J Med Arom Plant Sci 27(2):16–19

Somanath S, Bhaskar S, Khalak A (2004) Economics of cultivation of *Coleus forskohlii* as influenced by FYM, fertilizers and sources of potassium. J Root Cros 30(1):70–71

Srinivas NN, Patil GM, Alagundagi S (2008) Growth and yield of coleus intercropped in victory1 mulberry under integrated nutrient management. Karnataka J Agric Sci 2(4):524–526

Sudhakar DH (2005) Standardization of organic farming practices in coleus (*Coleus forskohlii* Benth). M. Sc. (Hortic.) Thesis. University of Agricultural Sciences, Bangalore

Sundharaiya K, Ponnuswami V, Jasmine AJ (2000) Effect of growth regulators in the propagation of sarkaraikolli (*Gymnema sylvestre*), medicinal coleus (*Coleus forskohlii*) and thippili (*Piper longum*). South Indian Hortic 48(1):172–174

Suresh MS, Koppad AG, Shivanna H (2010) Effect of spacing, manure and fertilizers on growth and yield of *Coleus forskohlii* under teak plantation. Karnataka J Agric Sci 23(2):397–399

Swamy KN, Rao SSR (2010) Effect of brassinosteroids on rooting and early vegetative growth of coleus (*Plectranthus forskohlii* (Willd.) Briq.) stem cuttings. Indian J Nat Prod Resour 1(1): 68–73

Tiwari RKS, Das K (2010) Sprouting and survival of *Coleus forskohlii* stem cuttings by hormonal pretreatment. J Med Aromatic Plant Sci 32(3):239–244

Veeraraghavathatham D, Venkatachalm R, Sundararajan S (1985) Performance of two varieties of *Coleus forskohlii* under different spacing levels. South Indian Hortic. 33:389–392

Veeraraghavathatham D, Venkatachalam R, Sundararajan S (1988) Effect of various levels of N, P and K on tuber yield of *Coleus forskohlii*. South Indian Hortic 36(5):252–257

Velmurugan M, Rajamani K, Paramaguru P, Gnanam R, Bapu JRK (2009) Studies on correlation and path analysis in mutants of coleus (*Coleus forskohlii* Briq.) for yield and forskolin content in V2M1 generation. J Horti Sci 4(1):63–67

Vennila C (2006). Evaluation of planting pattern and organics on *Coleus forskohlii* + *Phyllanthus amarus* intercropping system. PhD thesis. Tamil Nadu Agricultural University, Coimbatore

Vennila C, Jayanthi C, Nalini K (2008) Evaluation of planting pattern and nutrients on growth and yield of *Coleus forskohlii* intercropped with *Phyllanthus amarus*. Crop Res 36(1):359–361

Vennila C, Jayanthi K (2014a) Evaluation of productivity indices of *Coleus forskohlii* and *Phyllanthus amarus* intercropping system. Agric Sustain Develop 2(2):158–161

Vennila C, Jayanthi K (2014b) Effect of planting systems and sources of nutrients on productivity of medicinal coleus (*Coleus forskohlii*). Intern J Food Agric Veterinary Sci 4(2):97–101

Vennila C, Jayanthi K (2014c) Nitrogen use efficiency of medicinal coleus in relation to application of nutrients. Indian J Plant Sci 3(4):58–60

Vishwakarma RA, Tyagi BR, Ahmed B, Hussain A (1988) Variation in forskolin content in the roots of *Coleus forskohlii*. Planta Med 54(5):471–472

Yanagihara H, Sakata R, Shoyama Y, Murakami H (1995) Relationship between the content of forskolin and growth environment in clonally propagated *Coleus forskohlii* Briq. Biotronics 24: 1–6

Success Stories and Trade and Commerce of *Coleus forskohlii* 11

Abstract

Contract farming of *Coleus forskohlii* is gaining popularity among small and marginal farmers of Salem and Kanchipuram districts and neighborhood in India. In India, about 2500 tonnes of *C. forskohlii* are cultivated annually. The crop is widely grown in Karnataka, Tamil Nadu, and Gujarat. There are many success stories. Bangalore-based Sami-Sabinsa Corporation had brought about contract farming of *C. forskohlii* in Salem. They cultivated *C. forskohlii* across 10,000 acres in Salem, Villupuram, Vellore, and Thiruvannamalai districts in Tamil Nadu, India.

Keywords

Coleus forskohlii · Forskolin · Coleus farming · Success story · Sami-Sabinsa

11.1 Introduction

Contract farming of *Coleus forskohlii* is gaining popularity among small and marginal farmers of Salem and Kanchipuram districts and neighborhood in India. *C. forskohlii* contract farming is gaining popularity among the small and marginal farmers of Tamil Nadu. The crop is widely grown in Karnataka, Tamil Nadu, and Gujarat. "It is an easy and a profitable crop for the farmers," said Mr. G. Sivaji, an enterprising farmer of Navalur Therku Kadu village near Athur in Salem district. Encouraged by the growing popularity of traditional medicines like Ayurveda more farmers are opening up to the idea of cultivation of medicinal plants. The medicinal plants can be grown as an intercrop that provides additional income, but marketing is still a challenge.

Traditional farmers of Tirupur, Udumalpet, Dharmapuri, and Karuru increasingly raise medicinal herbs. T. Shaktivel, a farmer in Unathur, said that *C. forskohlii*

T. Pullaiah, *Forskolin*, https://doi.org/10.1007/978-981-19-6521-0_11

(*Marundhu Koorakannu* in Tamil), which he has been cultivating in 2.5 acres for 10 years was now more fetching him good returns.

There are many success stories. Bangalore-based Sami-Sabinsa Corporation had brought about contract farming of *C. forskohlii* in Salem. They cultivated *C. forskohlii* across 10,000 acres in Salem, Villupuram, Vellore, and Thiruvannamalai districts in Tamil Nadu, India.

11.2 Success Story—Dr. Muhammed Majeed

Dr. Muhammed Majeed is an internationally acclaimed scientist, entrepreneur, and pioneer in the field of evidence-based natural ingredients. He popularized the Indian Ayurvedic ingredients as food supplements in the USA and European countries. He established the company Sami-Sabinsa corporation. Today, Sami-Sabinsa Group, which currently employs over 700 people, is expecting a turnover of Rs 1250 crore in 2019–2020.

11.2.1 Sami-Sabinsa Corporation

The Sami-Sabinsa group pioneered the natural extracts of *C. forskohlii*, for use in nutritional and cosmetic applications, in the early 1990s. And the outcome was ForsLean®, a natural ingredient, forskolin-enriched extracts from the roots. In 1998, Sami-Sabinsa introduced a novel and previously unknown use for *C. forskohlii* (ForsLean®) for promotion of lean body mass in weight management. Sami Labs is primarily into cultivating *C. forskohlii* and its associate in the USA, Sabinsa Corporation discovered that *C. forskohlii* when consumed broke down fat without affecting other tissues. "Thus can be used as a good weight management product. In 1998, Sami Labs were granted a US patent for the discovery," said Majeed. ForsLean® is the extract of *C. forskohlii* (standardized to contain 95% forskolin) which is potentially useful in skin care formulations, particularly as a conditioning agent.

The company contracts with farmers for a range of herbals for its ingredients, including *C. forskohlii*. Sami Labs is benefiting more than 10,000 farmers commercially to grow *C. forskohlii* in Tamil Nadu and Karnataka by assuring them with a buyback guarantee (Fig. 11.1).

https://forslean.com/forslean/overview/

info@sabinsa.com

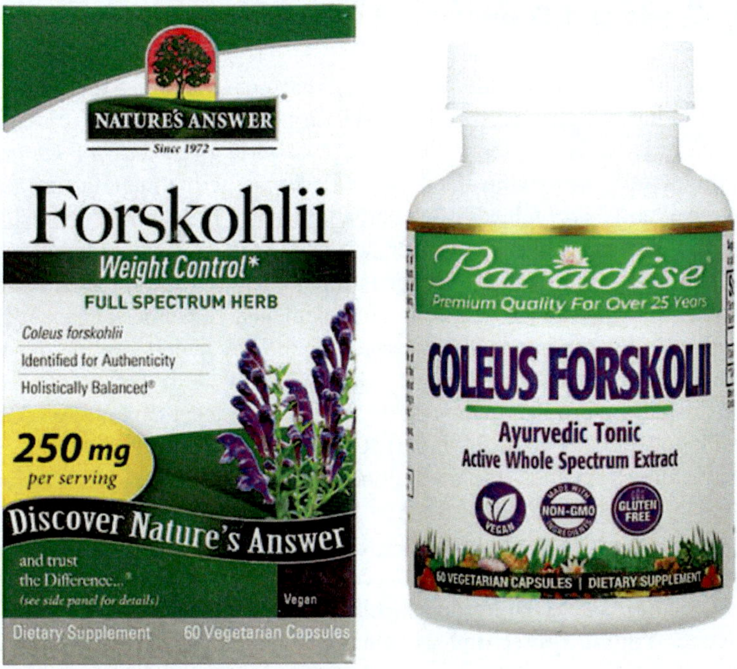

Fig. 11.1 Products of Sami-Sabinsa

Dr. Muhammad Majeed, Sabinsa corporation and Sami Labs Founder, in the field of *Coleus forskohlii* Image: Sabinsa

11.3 Trade and Commerce of *Coleus Forskohlii*

Estimated consumption by the domestic herbal manufacturing units, for the year 2005–2006, has been assessed at 1468 MT (dry wt.). During 2005–2006, exporters at Tuticorin (Tamil Nadu) reported export of 140 MT of *Coleus forskohlii*. Almost the entire quantity being consumed by the domestic herbal manufacturing units, as well as the exports, is obtained from cultivation. **The Coleus and Turmeric Market is expected to reach $6.9 billion by 2027, at a CAGR of 7.4% during the forecast period**. The growing demand for coleus forskolin in the sports nutrition industry, increasing consumption of weight loss supplements, consumer demand for natural products, and the growing Ayurveda sector are the key factors driving the growth of the coleus and turmeric market.

The key players profiled in the coleus market are Wacker Chemie AG (Germany), Ambe Phytoextracts Pvt. Ltd. (India), OmniActive Health Technologies Limited (India), Natural Remedies (India), Herbochem (India), Inventia Healthcare Limited (India), Sabinsa Corporation (U.S.), Star Hi Herbs Pvt. Ltd. (India), Arjuna Natural Pvt. Ltd. (India), Plant Lipids Private Limited (India), Pharmavit (Netherlands), Indena S.p.A. (Italy), Olive Lifesciences Pvt. Ltd. (India), Alchem International Pvt. Ltd. (India), and Sanat Products Ltd. (India).

Shaanxi Yougu Biotechnology Co., Ltd., established in **2010** at Xian in Shaanxi, is a leading exporter, manufacturer, and supplier of plant extract in China. Shaanxi Yougu Biotechnology Co., Ltd. is one of Trade India's verified sellers of listed products. They are exporting, manufacturing, and supplying *Coleus forskohlii* extract forskolin to their customers from Xian, Shaanxi, China.

Bionatural 21, Suneja House, first Floor, Gulmohar Enclave Community center, New Delhi –110,016 info@bionatural.inhttps://www.bionatural.in/product/coleus
***Coleus forskohlii* Grade available:** Forskolin 10% & 20%.

Shining Star Exports, Chennai in Tamil Nadu, is a leading Exporter, Manufacturer, Supplier, and Trading Company of Herbal & Botanical Products in India

CK AND CO, Salem, Tamil Nadu

Gtr Overseas, Tiruchirappalli, Tamil Nadu

Cymbio Pharma Private Limited, Yelahanka, Bengaluru, Karnataka

Shubhasya Biotech, BTM Layout, Bengaluru, Karnataka

Arizone International LLP, Established in **2014** at Daman in Daman and Diu, is a leading Exporter, Manufacturer, Supplier, and Trading Company in India.

References

Kannan RR, Rohini AK, Angles S, Murugananthi D (2020) An economic analysis of coleus production and marketing in Tamil Nadu. International Journal of Education Management 10(1):18–22

http://www.agritech.tnau.ac.in/success_stories/sucess%20stories_Hoticulture.html#contract2
https://www.seair.co.in/coleus-forskohlii-export-data.aspx
http://www.dmpharma.co.in/Coleus%20Forskohlii%20Root%20Extract%20soft%20gelatin%20Capsule.html
http://timesofindia.indiatimes.com/articleshow/57283504.cms?utm_source=contentofinterest&utm_medium=text&utm_campaign=cppst